A

Cytodifferentiation
and
Macromolecular Synthesis

The Twenty-First Symposium
The Society for the Study of
Development and Growth

Asilomar, California, June 1962

Cytodifferentiation
and
Macromolecular Synthesis

Edited by

Michael Locke

Developmental Biology Center
Western Reserve University
Cleveland, Ohio

1963

ACADEMIC PRESS, New York and London

ACADEMIC PRESS INC.
111 Fifth Avenue, New York 3, New York

United Kingdom Edition published by
ACADEMIC PRESS INC. (LONDON) LTD.
Berkeley Square House, London W.1

LIBRARY OF CONGRESS CATALOG CARD NUMBER: 63-14493

First Printing, 1963
Second Printing, 1964

PRINTED IN THE UNITED STATES OF AMERICA

List of Contributors

Numbers in parentheses indicate the page on which the author's contribution begins.

JOSEPH G. GALL, Department of Zoology, University of Minnesota, Minneapolis, Minnesota. (119)

S. GRANICK, The Rockefeller Institute, New York, New York. (144)

PAUL B. GREEN, Division of Biology, University of Pennsylvania, Philadelphia, Pennsylvania. (203)

CLIFFORD GROBSTEIN, Department of Biological Sciences, Stanford University, Stanford, California. (1)

JEROME GROSS, Department of Medicine, Harvard Medical School, Massachusetts General Hospital, Boston, Massachusetts. (175)

HEINZ HERRMANN, Institute of Cellular Biology, University of Connecticut, Storrs, Connecticut. (85)

FRANÇOIS JACOB, Services de Génétique microbienne et de Biochimie cellulaire, Institut Pasteur, Paris. (30)

CHARLES M. LAPIERE, Department of Medicine, Harvard Medical School, Massachusetts General Hospital, Boston, Massachusetts. (175)

JAMES W. LASH, Department of Anatomy, School of Medicine, University of Pennsylvania, Philadelphia, Pennsylvania. (235)

CLEMENT L. MARKERT, Department of Biology, The Johns Hopkins University, Baltimore, Maryland. (65)

JACQUES MONOD, Services de Génétique microbienne et de Biochimie cellulaire, Institut Pasteur, Paris. (30)

MARVIN L. TANZER, Department of Medicine, Harvard Medical School, Massachusetts General Hospital, Boston, Massachusetts. (175)

CHARLES YANOFSKY, Stanford University, Stanford, California. (15)

Foreword

"...Our Microscope will easily inform us that the whole mass consists of an infinite company of small Boxes or Bladders of Air...

"...I told several...of these small Cells placed endways in the eighteenth part of an inch, whence I concluded there must be...in a Cubick Inch about twelve hundred Millions...a thing almost incredible, did not our microscope assure us of it by ocular demonstration;...so prodigiously curious are the works of Nature, that even these conspicuous pores of bodies...are yet so exceeding small, that the Atoms which Epicurus fancy'd would go neer to prove too bigg to enter them, much more to constitute a fluid body in them...

"...Now, though I have with great diligence endeavoured to find whether there be any such thing in those Microscopical pores... Yet have I not hitherto been able to say anything positive in it; though, methinks, it seems very probable, that Nature has in these passages, as well as in those of Animal bodies, very many appropriated Instruments and contrivances, whereby to bring her designs and end to pass, which 'tis not improbable, but that some diligent Observer, if help'd with better Microscopes, may in time detect..."

Observ. XVIII. Of the Schematisme or Texture of Cork,
and of the Cells and Pores of some other such frothy Bodies.
Robert Hook, 1665

At Asilomar in June, 1962, a group of latter day diligent observers reported progress to the Society in their studies on the means whereby nature brings her designs and ends to pass, helped by the National Science Foundation and the local committee, to whom all thanks.

M.L.

Contents

Quantitative Studies of Protein Synthesis in Some Embryonic Tissues

Cytodifferentiation and Macromolecular Synthesis

CLIFFORD GROBSTEIN

Department of Biological Sciences, Stanford University, Stanford, California

The title of this essay, and this volume, was the theme of the 21st Growth Symposium. The number of the symposium may be significant, for there is a sense in which the Growth Symposium, its sponsoring Society, and the science of development which is the Society's focus, all have been coming of age. After a long period of successful operation provided by the impetus of its founders, the Society has been re-evaluating and redefining its role. The appearance of this volume, under a new publisher and a new editor, comes only after careful soul-searching by the Society's Publication Committee. Despite altered circumstance since the Growth Symposium was founded, it was decided that its publication remains a valuable contribution to the literature of developmental biology, and that it is the responsibility of the Society to make it an even greater one. To accomplish this, changes—involving both procedures and content—were suggested. Among these was looser "coupling" of the orally presented symposium and the published one, and inclusion of a synthetic article on the theme of the symposium.

Traditionally, the President of the Society for the Study of Development and Growth, in consultation with the Executive Committee, has been the chief organizer of the symposium. Having suggested the theme it is reasonable that the President should comment on it, or should invite some one else to do so whose views may prove illuminating. The transitional status of the 21st Symposium, planned during the deliberations of the Publications Committee, provided no opportunity to select an alternative commentator. Hence, responsibility fell to the President and this essay was written after the symposium had occurred and with full knowledge of its content.

It is no disparagement of speakers to note that symposium organizers frequently find their chosen theme honored, if at all, more in the introduction than in the body or conclusions of the papers. This seems inevitable, since the speakers do not necessarily hold the same conception of the theme as the organizer, and sometimes are dragged in, bitterly protesting, under its umbrella. Occasionally a speaker sees the organizer as tyrannical, and the

organizer sees the speaker as recalcitrant and resentful in the face of integrating guidance. For these and other reasons, the preconceived theme may
be less sharply delineated than hoped. The present symposium is no exception; its papers as a set clearly fit the theme, but the treatment is neither as
focused nor as complete as might be provided by a single author pursuing
a single conception. In this lies opportunity and justification for an effort
at synthesis. To restate the theme as conceived, and to round out the treatment by the authors from the point of view of the organizer—these are the
objectives which, if they accomplish little else, at least provide opportunity
for a last word for the organizer.

The theme may be rephrased as a hypothesis, which indeed it is. A number of authors in recent years (e.g., Spiegelman, 1948; Markert, 1956, 1960)
have suggested that the differentiation of cells is fundamentally a switching
of biosynthetic activity, leading to the appearance of new macromolecular
species whose accumulation or export is manifested as specialized structure
or activity. Whether or not differentiation is solely or primarily such switching of biosynthesis may be questioned (Ebert, 1955), but certainly it is involved in many instances. In the context of spectacular advances in knowledge of biosynthesis, particularly of proteins, the hypothesis has been given
new content. Biosynthesis of proteins is now firmly linked to the gene at one
end, and to cytoplasmic assembly centers somewhere near the other end.
Between the fixed genotypic nucleotide sequence of deoxyribonucleic acid
(DNA), and the alterable phenotypic expression in a characteristic protein,
lie messenger ribonucleic acid (RNA), amino acid activation, the yet to be
clarified order within the ribosome, and processes of polymerization, folding,
and complexing of macromolecules whose importance increasingly is emphasized. If a new molecular species is to appear, whether *de novo* or through
increase in amount, some alteration must be effected along this chain from
gene to final product. Induction of differentiation thus would be effected
through control exerted along the chain of biosynthesis. With the relative
paucity of information about control of intermediate steps in the chain, particularly in complex organisms which most characteristically differentiate,
it has been natural to concentrate attention on the point of most precise information, on the relationship between gene and biosynthesis. Yanofsky in
his chapter in this volume reviews some of the exciting new information in
this area, indicating that amino acid sequence of particular proteins is coded
in, and ultimately determined by, nucleotide sequence in the DNA of the
gene. Given the genetic conception of replicating nucleotide sequences, and
the assumption of a system for transcribing these into sequences of other
building blocks such as amino acids, cytodifferentiation can be, and has

been, visualized as the switching on and off of synthesis at particular genetic sites (see chapter by Markert). Precedent for this exists in microbial systems, and Jacob and Monod in their paper present a number of models based on enzyme induction, which could operate as controls. What is important to the general hypothesis is not so much the details of the models, which can be adjusted to yield anything desired, but the provision of regulation, i.e., alteration of genomic product without alteration of nucleotide sequence. The genome is assumed to undergo "functional" change, without sacrifice or modification of properties detected by allowing it to replicate. Differentiative macromolecular synthesis thus stems from control of the functional or transcriptive behavior of the genome, without alteration of its genetic behavior as classically conceived in breeding experiments.

This is a provocative and powerful hypothesis and, of course, deals perfectly with what has been called the dilemma of differentiation, how nuclei, classically assumed to be genetically equivalent, can be controlling cells so different as neuron, macrophage, and melanocyte. This so-called dilemma, man-made rather than cell-made, has never been as sharp or fearsome as sometimes has been portrayed. It has several possible resolutions not yet excluded by the evidence, viz., the nuclei of differentiated cells may not remain genetically equivalent, some extranuclear genetic change may occur in differentiation, or differentiation may not involve a genetic change at all. The first possibility has been raised seriously by nuclear transplantation studies (Briggs and King, 1959), the second by data suggesting genetic stability in cytoplasmic particles (see chapter by Granick), the third by challenge to the assumption that differentiated states are replicatively transmitted through successive cell divisions (Trinkaus, 1956; Grobstein, 1959). The Jacob and Monod model, applying primarily the newer understandings of high-resolution microbial genetics, dissolves the dilemma in yet another way. It distinguishes two kinds of genetic mechanism, that involved in replication of the code and that governing its transcription. Nuclei may have a complete set of replicative sites, but be transcribing only from particular ones at particular times. Moreover, and here the model advances beyond earlier speculations and reaches out to link genetic and physiological approaches to cell function, transcription of structure-determining sites of synthesis may be controlled by other replicative sites (regulators) through the intervention of epigenetic inducers and repressors. By defining *genetic* to include both replicative and transcriptive operations, nuclei simultaneously may be genetically equivalent (when tested by breeding behavior for replicative properties) and genetically nonequivalent (when tested by synthetic behavior for metabolic properties).

This interesting resolution has an additional feature of possible significance. It has long been recognized that most differentiations in higher organisms involve more than a single new biosynthetic pathway; that a given cell type must be defined in terms of a constellation of syntheses and properties. The concept of the operon, of transcriptively coupled syntheses turned off and on by a single operator gene, conceivably in response to a single external intervention, may prove helpful in this connection. Compounding simplifications, one might think of the number of differentiated cell types in a higher organism as some function of the number of operons in its genotype.

Thus, the hypothesis of macromolecular synthesis as the key to cytodifferentiation has acquired a codicil, close impingement of genetic control. This increases its interest and attractiveness in relation to a general theory of cell heredity and function. It does not make easier, however, its evaluation as a specific theory of cytodifferentiation, for very few data currently are available to test critically the specifically genetic aspect. The closest applicable information comes from the remarkable cytological observations, summarized by Gall in his article, which are correlating details of chromosome structure with cell type. The evidence for localized synthetic activity along the length of the chromosome, in patterns characteristic for a differentiated cell type, clearly conforms with the hypothesis. Without greater resolution, however, whether chemical, genetic, or morphological, it cannot yet be said that what are observed are actual functioning genetic sites. Nor is anything known of the details of control of these sites. It is not clear whether the localized changes are first steps leading to differentiation, or early consequences of differentiation initiated elsewhere. Certainly nothing can be said as to the possible involvement of regulator genes, or of operators. Nonetheless, there now has emerged out of an impressive array of genetic, biochemical, cytological, and embryological data, a working hypothesis of cytodifferentiation at the molecular level. Particularly in the genetic version of Jacob and Monod, it is based largely on microbial data and its applicability to multicellular systems— whose behavior is more traditionally and spectacularly differentiative—is only beginning to be tested. It seems worthwhile to evaluate and comment on the applicability of the working hypothesis to embryonic differentiation, both to emphasize implications for future work and to reduce terminological and other differences between the several parties of investigators who are now converging from different directions on a common problem area.

The question of definition of differentiation deserves brief consideration. What is under discussion, of course, is not differentiation of the whole organism but cytodifferentiation, i.e., the cell changes which accompany and reflect the increasing heterogeneity which is one of the hallmarks of development in complex organisms. The changes are in the direction of cell speciali-zation—concentration on certain activities and syntheses at the expense of others. Jacob and Monod propose to define differentiation in a manner which is entirely appropriate for microbes, viz., "Two cells are differentiated with respect to each other if, while they harbor the same genome, the pattern of proteins which they synthesize is different." Several points should be kept in mind, however, in judging applicability to higher organisms.

First, in multicellular systems the only differentiations demonstrably ad-missible under a standard of genomic equivalence are those of gametogenesis. It is often overlooked that in gametogenesis cytodifferentiations of extreme degree occur without alterations of the genome in the replicative sense. Taken by itself gametogenesis strongly implies that specialized syntheses *can* proceed without alteration of nucleotide sequence. Apart from this case, however, a flat assumption of genomic equivalence will seem to some to beg the question for multicellular organisms. Hope that it may be possible to test the genome of differentiated cells is higher than seemed justified some years ago, but it has not yet been done unequivocally. Nonetheless, the new insights into the nature of genetic coding, and the discrimination of a separately regulated process of transcription, are widely felt to make message changes less attrac-tive than reading changes as a mechanism of differentiation. Pending evidence to the contrary, and since it constitutes a basis for a working hypothesis of considerable heuristic value, it seems reasonable to accept genomic equiva-lence provisionally.

Second, in multicellular systems (as probably in microbial) acceptable differentiative changes are not limited to biosynthesis of proteins. Behavioral changes such as motility and phagocytosis, as well as synthesis of complex polysaccharides, lipids, or steroids, are distinguishing criteria of certain dif-ferentiations. It is not unreasonable to assume, however, that underlying these is the synthesis of specific proteins if only in catalytic amounts. Again as a working hypothesis, the appearance of a new pattern of synthesis of pro-teins seems a reasonable criterion of differentiation.

Third, a number of cyclical changes in cellular synthesis, illustrated by hormonally controlled changes in secondary sex structures, have not usually been regarded as differentiative. The term modulation has been applied to such changes when it is clear that they are without stability in the absence of the initiating circumstance. However, the borderline between these changes and "true" differentiations has never been sharp, and there seems to be no good reason why they should not be grouped under the hypothesis as differentiations of minimal intrinsic stabilization.

In view of these considerations the proposed definition of Jacob and Monod perhaps may be rephrased as follows. There is a class of phenomena involving change of some stability in the cellular pattern of protein synthesis, without known change in the replicative properties of the genome. The class includes at least several subclasses which may have mechanisms in common: bacterial induction-repression, mating types in *Paramecium* (Sonneborn, 1960), and embryonic cytodifferentiation. In applying the term differentiation to the class, one emphasizes the likelihood of important similarities in the subclasses without intending to minimize the equally important differences which may exist as well.

More interesting than definition, however, is the question whether properties implied by the model are common to the subclasses. For example, is it the case that embronic differentiation appears to involve genetic transcription? A whole series of examples, both classical and recent, answer affirmatively. The results of heteroplastic and xenoplastic developmental interactions, together with the occurrence of mutations affecting the structure of such specialized proteins as hemoglobin, show incontrovertibly that genetic sites are operative during differentiation, and hence must be coupled in some fashion with its controls.

More difficult to analyze is the specific implication that differentiative controls impinge on genetic sites through combination with repressors of structural genes. The matter has two aspects, whether repressors of structural genes actually exist in cells of higher organisms, and, if so, whether factors known to control embryonic cytodifferentiation interact with such elements of the genetic system. Concerning the first aspect there is as yet only scanty indirect evidence, which has been referred to elsewhere (Jacob and Monod, 1961) and to which nothing can be added here. Concerning the second aspect there is more evidence, enough to put the matter in perspective, though not enough to draw firm conclusions. The evidence comes from the study of developmental induction, in the course of which new differentiative pathways are established. Since the discovery that differentiative processes in one tissue are controlled through intimate association with another, embryologists have wrestled with the nature of the mechanisms involved. Though these mechanisms cannot yet be specified, the current status of the problem has two implications for the Jacob-Monod model: it suggests that dependent differentiation as seen in developmentally coupled tissues is favorable test material, and it warns that there may be several levels of control between an inducer as observed in embryos and the primary effector, F, postulated for microbial genetic systems.

With respect to favorable test material, it is clear that one would like a population of cells, as large and homogeneous in behavior as possible, which can be controlled developmentally so as to initiate synthesis of well-characterized specialized proteins under defined nutritional conditions. It would be useful if genetic variants affecting the character of the proteins were available or could be readily produced. No such ideal system has yet been described, but a number of laboratories have these requirements in mind and advancing tissue culture technology is providing promising candidate-systems. Particularly attractive are the differentiation of muscle, cartilage, melanocytes, erythrocytes, fibroblasts, lens, and various epithelial cells. In our laboratory we are especially impressed by the potentialities of epithelio-mesenchymal rudiments, where presence or absence of mesenchyme provides control over epithelial differentiation and, in such rudiments as the pancreas, the epithelium produces a number of characterizable proteins.

Investigation of such systems, along with much other data on embryonic induction *in vivo* and *in vitro*, has not encouraged the view that differentiative control of this kind involves direct impingement of products of one tissue on the genetic system of another. In discussing this we may treat the regulator-operator-structural gene complex of Jacob and Monod as a component (black box), and ask whether its input seems likely to be a molecular species emitted by, and coming directly from, a second tissue acting as an embryonic inductive source. Or asked in another way, does it seem likely that embryonic induction involves a single step, the direct initiation of synthesis of differentiated product?

In my own view this does not seem likely, though the possibility is not excluded for some instances. I am impressed by the following considerations:

First, in many inductions the response involves the appearance of several cell types, and these in a complex temporal and spatial pattern. Primary induction by chorda mesoderm is a good example, but many secondary inductions are simpler only in degree, e.g., the response of metanephrogenic mesenchyme to ureteric bud involves appearance of glomeruli and secretory tubules containing a number of cell types. It is hard to see how the inducer could directly turn on biosyntheses of so many different sorts in so many cell types. Either a battery of molecular species, or intermediate relays, seem likely.

Second, some inductive responses can be initiated by a number of chemically unrelated molecular species. Hence, specific chemical information does not appear to be required to be fed in by the inducer. Allosteric induction, as emphasized by Jacob and Monod, provides a degree of freedom in the regulation of microbial biosynthesis. In embryonic induction there appears to be even less coupling of effector and response, implying the existence of yet additional intermediate steps.

Third, the kinetics of embryonic inductive processes, where they have been examined, frequently do not suggest a single step. In kidney tubule induction, for example, exposure of some 30 hours to inducer is required before the process will continue independently. In cartilage induction the minimum exposure time is shorter, less than 12 hours, but the characteristic product is not detected until 3–4 days. The long latent period does not seem to accord with a direct derepression of chondroitinsulfate synthesis, and Lash's report in his chapter of early sulfate binding in some other form may be significant in illuminating intermediate reactions.

Fourth, during the latent period before the appearance of a characteristic product, other changes occur, particularly in the shape and orientation of cells with respect to the interface of interaction.

Fifth, in addition to the effect of cells of unlike type (heterotypic induction), there is evidence that cells of like type may affect each other's differentiation (homotypic or homeogenetic induction). The relation between heterotypic and homotypic controls is not clear, but there is some evidence that the former may operate through the latter and hence not directly on the genetic system of the cell (Grobstein, 1962a).

No one of these considerations is so cogent as to exclude direct intervention of certain inducers in the genetic system of the responding cell. More incisive studies that bear on these matters are needed and may be expected soon to be forthcoming. For the moment, however, the combined information suggests that the effector, F, which is disarmingly simple in the model of Jacob and Monod, itself probably has a considerable control history before becoming available to combine with the repressor. Some of the possibilities are suggested in the modified model in Fig. 1 where regulator and operon are located within the chromosome, giving rise extrachromosomally to repressor and messenger. The extrachromosomal control circuit postulated by Jacob and Monod runs from messenger, through synthetic unit, through synthetic product (here labeled monomer), and back to repressor. The key step from the point of view of embryonic induction is the combination of effector and repressor, and the key question is the source and nature of the effector.

It may be the synthetic product itself, acting in any of the various ways diagrammed by Jacob and Monod (1). To convert these circuits into mechanisms of induction, one need only assume transfer of synthetic product from cell to cell (1A). But it also may be any one of a number of other possibilities which have been put forward over the years to explain inductive mechanisms. For example, the effector may exist in the pre-induced cell, but needs to be released by the inducer from an inactivating complex (2). The inducer may have to interact with a synthetic product in order to produce an effector (3),

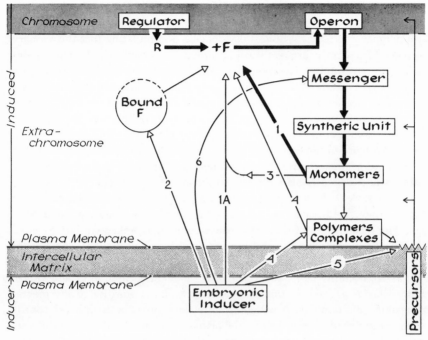

FIG. 1.

or it may affect polymerization and complexing of the product, leading perhaps to surface and matrix changes whose consequences include effector availability (4) or precursor availability (5). Or inducer may operate at the level of messenger (6), itself acting as messenger or altering the stability and effectiveness of intrinsic messenger.

Combinations of these and other possibilities can provide for an embryonic inducer, implying differentiative control, to act at any of a series of successively more peripheral circuits, proceeding outward from chromosome, to nucleus, to cytoplasm, to cell surface, to associated cells of like and unlike type. The only requirement imposed by the model is that any control device, at whatever shell of influence it operates, shall funnel down in its effects on *F*, the critical switch for the genetic component. This emphasizes that embryonic inducer and microbial inducer may not be assumed to be equivalent; rather their relationship is one of the questions posed by the model. It emphasizes, further, that the Jacob and Monod model focuses on only one area of a larger model of cytodifferentiation as controlled biosynthesis, on the nature of the genetic component and its link with other circuits of control.

It should not be forgotten, in this connection, that the production of defined amino acid sequences does not exhaust the developmentally significant path of macromolecular synthesis. Evidence is accumulating that the primary sequences frequently are variously coupled as subunits in larger physiologically active units. Recent information on lactic dehydrogenase (see chapter by Markert) and the collagen story, with its demonstration of controlled polymerization and complexing as the basis of alteration of ultrastructure (see paper by Gross), are excellent cases in point. Not only do such secondary and tertiary variations themselves constitute differentiation, but there is evidence in the case of collagen that they provide the basis for further differentiation in two ways. Through their steric properties they control the addition of new materials such as hydroxyapatite, of crucial importance to bone and skeletal formation. In addition, through their own maturational changes they "feed back" to control cellular behavior including differentiation. The mechanisms of these effects are completely unknown, but together with studies of cell adhesion and movement they raise questions whether the cell periphery does not have important control functions in biosynthesis, not only as an assembly point at levels higher than the formation of amino acid chains, but as a gateway to synthetic sites which is sensitive to the macromolecular environment. Nowhere is this question of a peripheral control center more sharply posed than in the primary step to all subsequent differentiation, the cortical activation accompanying fertilization. Here it is established beyond question that peripheral change triggers biosynthetic initiation.

To emphasize these possibilities is not to urge complication of the model for complexity's sake. The impression of need for successive shells of control in differentiation arises from the results of developmental analysis. These results have not, as so spectacularly in the case of recent genetic analysis, achieved a conceptual breakthrough. Nonetheless, the data which have been accumulating suggest to the perceptive what is to come, and models which fail to take them into account are accordingly diminished in utility. In particular instances there is evidence that protein synthesis, whether directly or indirectly, can be turned on and off by hormones (e.g., ecdysone), substrates (phenylalanine), antibodies, chemically uncharacterized intercellular materials, interactions between cells of like type and between cells of unlike type. It is not impossible that all of these involve direct provision of an F-substance. It seems more likely, however, that some may operate through others and that developmental integration takes advantage of control possibilities available at several levels. There is a rationale for this beyond the immediate context. For example, not only would such an arrangement protect the genetic memory stream of past successful adaptation against disturbance or loss, but it would extend the range of adaptation of any given genotypic structure. A nucleotide sequence which had, so to speak, to make its own way in the world could survive and replicate only in physicochemical environments conducive for the particular nucleic acid. A nucleotide sequence wrapped in a translated, insulating shell may increase its range of adaptation by the properties of the shell. With each successively added shell, insulation and range may increase, but so do the complexities of regulation and control. In these terms, the evolutionary trend to higher levels of organization is the concomitant of ever-expanding occupancy of the environmental spectrum, and development is fluctuation between the minimum number of shells necessary for transmission of the memory stream between generations, and the optimum number for particular adapted states. It would seem remarkable, under these circumstances, if all outer regulations impinged directly on the innermost, or if the innermost directly controlled the outermost. Translation through successive intermediates would much amplify the variant forms in which a given nucleotide sequence change could be tested for evolutionary survival.

All of these considerations argue that the conception of cytodifferentiation as controlled biosynthesis does not stand or fall with the assumption of direct control at the genetic level, intriguing though the microbial data make this possibility appear and efficient though such simple switching may seem from the point of view of a cell-machine (see chapter by Markert). Setting this question aside, however, there are other features of the conception which have important implications for future studies of differentiation. Not least among these is the "handle" immediately provided for the quantitative definition and measurement of the differentiative process. Quantitation of differentiation has been a difficulty in the past because each instance has been identified with a process or product peculiar to it, and hence the criteria used have been incommensuarable for comparison among instances. Viewing differentiation as specialized macromolecular synthesis offers a hopeful approach to this problem. For example, the ratio between the rates of synthesis, or of the accumulated amounts, of differentiative proteins and generalized proteins may provide a useful index of degree of differentiation, both in expressing the time course of a particular differentiation and in comparing different courses. Beyond this, one can perhaps express the level of differentiation in terms of each of the several syntheses which characterize a cell type, e.g., chymotrypsin, trypsin, and ribonuclease for a pancreatic acinar cell, so as to see how closely coupled they are in their control. It would be interesting to know, for example, whether there are clusters of enzymes which appear simultaneously and which are commonly affected by genetic change, as might be expected if the properties of a particular differentiation result from control by a single operon. The sensitivity of detection of differentiation may be heightened, not only by identifying a characteristic product in smaller quantities as already has been done, but by recognizing such essential preliminary steps as transfer of messenger from chromosome to synthetic sites. In this connection it is worth keeping in mind that differentiation may provide special opportunities for the study of intermediate and transitory control steps in biosynthesis, if it is indeed a time of massive qualitative change. There are obvious technical problems in conducting such analyses, distinguishing reliably, for example, between synthesis and accumulation of product, but these can be expected to yield to concentrated ingenuity.

Not least among the advantages of viewing cytodifferentiation as control of macromolecular synthesis is the flexibility it provides with respect to the stabilization of differentiative change. Certainly there can be few subjects on which the ratio of discussion to cogent facts is so high as on this one. There has been a strong tendency to regard differentiation as a process *sui generis*, lying between mutation on the one hand and functional change on the other, and sharply set off from each in mechanism. This was most clearly expressed in the assumptions that differentiation is a replicative change of cell properties not involving genetic alteration, and that cell changes which are not stable in this sense are not truly differentiative. The models now under consideration require no such assumptions. They allow alteration of biosynthesis to occur in a number of ways, some replicative and classically genetic, others replicative but only quasi-genetic, still others of nonreplicative stability grading down to "mere functional change." It is clear that proper manipulation of the controls (assumptions) can provide all degrees from transitory functional to stable, quasi-genetic modification of biosynthetic activity. The models require no sharp cleavage between function and differentiative change, rather they allow the one to lead into the other.

This is advantageous because it allows the matter of stabilization to be approached as a question to be investigated rather than as a canon *ex cathedra*. In this regard, it is worth recalling that the evidence supporting the assumption that embryonic cytodifferentiation invariably is a fully stable and replicative *cell* state has been seriously challenged in several quarters. Recent discussions emphasizing replicability as a requirement of models of differentiation seem to disregard this. Jacob and Monod, in speaking of embryonic differentiation, remark that "this diversity is clonally transmitted in a stable way." Davis (1961) observes that differentiation "aims at bringing about differences among cells that will be maintained in each cell and its progeny." As has been pointed out elsewhere (Grobstein, 1959, 1962b), the earlier tissue culture literature, usually cited in this connection, is less than convincing in demonstrating that the differentiated states which arise during development are transmitted through cell generations in the absence of conceivably stabilizing interactions within cell groups. The more recent tissue culture studies, which have focused on the question of stability and propagability of differentiative biosynthesis at the cell level, emphasize the difficulty of demonstrating continuance of specialized synthesis after relatively short periods of culture. There is increasing suspicion, indeed, that in a given cell specialized synthesis and preparation for division are competitive and may, in some instances, be exclusive. More critical information is urgently needed, but meanwhile stabilization through cell generations does not appear to be a requirement of a general model of cytodifferentiation. Some stabilization of differentiated states clearly occurs, and mechanisms of stabilization must be provided in any successful model. The mechanisms may lie partly in the genetic system, now more broadly defined to include such components as messenger RNA which may prove more stable in cells of higher organisms than it appears to be in bacteria. Beyond this, however, the evidence suggests that stabilization mechanisms probably exist which are not replicative in the direct sense but which involve control circuits extending beyond the cell boundary, and even beyond the limits of populations of like metabolic type.

Indeed, it does not seem superfluous to point out that differentiative controls probably are a sector of organismal regulatory mechanisms which themselves have an ontogenetic history. Diversification of cells is the focus in cytodifferentiation; its essential concomitant from the point of view of the whole organism must be alteration of an integrating milieu. The power and attractiveness of the model of cytodifferentiation as macromolecular synthesis lie in its promise of unification, not only of the molecular, the genetic, and the developmental, but of micro- and macroregulation in embryo and adult as well. In the evolutionary progress from microregulation at the level of the genetic component (with a primary requirement for stability) to the macroregulation of complex adult organisms (with a primary requirement for adaptability), the rise of successive, more peripheral shells of control is clear. It would be surprising, in these terms, if specialized biosynthesis were solely under "primitive" microcontrol. The successive addition of peripheral control systems of greater exosensitivity, a kind of recapitulation of regulatory systems comparable to that occurring in functional and morphological spheres, seems more likely. It will be interesting to see, as experiments on differentiating systems are designed to test the provocative model of cytodifferentiation as macromolecular synthesis, whether this proves to be the case.

REFERENCES

BRIGGS, R., AND KING, T. J. (1959). Nucleocytoplasmic interactions in eggs and embryos. *In* "The Cell" (J. Brachet and A. E. Mirsky, eds.), Vol. 1, pp. 537–618. Academic Press, New York.

DAVIS, B. D. (1961). Opening address: The teleonomic significance of biosynthetic control mechanisms. *Cold Spring Harbor Symposia Quant. Biol.* **26,** 1–10.

EBERT, J. D. (1955). Some aspects of protein biosynthesis in development. *In* "Aspects of Synthesis and Order in Growth," Growth Symposium No. 13 (D. Rudnick, ed.), pp. 69–112. Princeton Univ. Press, Princeton, New Jersey.

GROBSTEIN, C. (1959). Differentiation of vertebrate cells. *In* "The Cell" (J. Brachet and A. E. Mirsky, eds.), Vol. 1, pp. 437–496. Academic Press, New York.

GROBSTEIN, C. (1962a). Interactive processes in cytodifferentiation. *J. Cellular Comp. Physiol.* **60,** Suppl. 1, 35–48.

GROBSTEIN, C. (1962b). Differentiation: Environmental factors, chemical and cellular. *In* "The Biology of Cells and Tissues in Culture" (E. N. Willmer, ed.). Academic Press, New York, in press.

JACOB, F., AND MONOD, J. (1961). On the regulation of gene activity. *Cold Spring Harbor Symposia Quant. Biol.* **26,** 193–211.

MARKERT, C. L. (1956). The ontogeny of divergent metabolic patterns in cells of identical genotype. *Cold Spring Harbor Symposia Quant. Biol.* **21,** 339–348.

MARKERT, C. L. (1960). Biochemical embryology and genetics. *Natl. Cancer Inst. Monograph* **2,** 3–18.

SONNEBORN, T. M. (1960). The gene and cell differentiation. *Proc. Natl. Acad. Sci. U. S.* **46,** 149–165.

SPIEGELMAN, S. (1948). Differentiation as the controlled production of unique enzymatic patterns. *Symposia Soc. Exptl. Biol.* **2,** 286–325.

TRINKAUS, J. P. (1956). The differentiation of tissue cells. *Am. Naturalist* **90,** 273–288.

Genetic Control of Protein Structure

CHARLES YANOFSKY

Stanford University, Stanford, California

Basic to any attempt at elucidating the mechanisms of regulation of gene activities is the need for an understanding of the relationship between gene structure and protein structure. Studies of gene structure and biological activity and of hereditary abnormalities in human hemoglobins and other proteins have provided convincing evidence for a direct relationship between gene and protein (see Fincham, 1960; Yanofsky and St. Lawrence, 1960). The simplest working hypothesis consistent with the findings of these studies is that the amino acid sequence of each protein reflects the linear sequence of the nucleotide-containing coding units of a gene. This hypothesis and many other aspects of the structural relationships between gene and protein are as yet untested. The work I will describe is being performed in the hope of obtaining information on these questions. The findings that we have obtained to this date bear on three aspects of this relationship, the genetic map in relation to protein structure, amino acid substitutions and the genetic code, and the translation of nucleotide sequences into amino acid sequences.

Characteristics of the Gene-Protein System Selected for Study

All of our work has been carried out with the tryptophan synthetase enzyme system in the bacterium *Escherichia coli* (Yanofsky, 1960). This enzyme system is required for the biosynthesis of tryptophan and catalyzes the three reactions shown in Table I. The third reaction is believed to be the physiologically essential reaction in tryptophan formation. The tryptophan synthetase system of *E. coli* is a somewhat unusual enzyme system in that it consists of two separable protein components (Crawford and Yanofsky, 1958). These components have been designated the A and B proteins. As can be seen in Table I, the AB complex catalyzes all three reactions while the A and B proteins separately are each active in only one reaction. Furthermore, the A protein is only 1% as active as the AB complex in reaction 2 and the B protein is only 3% as active as the AB complex in reaction 1. Neither protein by itself is active in the physiologically important reaction, reaction 3. Of the two proteins, the one that we have selected for the study of gene-protein rela-

TABLE I

THE REACTIONS CATALYZED BY TRYPTOPHAN SYNTHETASE

Reaction	Catalyzed by
(1) Indole $+$ L-serine $\xrightarrow{\text{B}_6\text{P}^a}$ L-tryptophan	AB; B
(2) Indoleglycerolphosphate $\rightleftharpoons$ indole $+$ glyceraldehyde-3-phosphate	AB; A
(3) Indoleglycerolphosphate $+$ L-serine $\xrightarrow{\text{B}_6\text{P}^a}$ L-tryptophan $+$ glyceraldehyde-3-phosphate	AB

^{a} B_6P = pyridoxal phosphate.

tionships is the A protein. This protein is the smaller of the two (molecular weight = 29,500) (Crawford, 1962; Henning *et al.*, 1962), it is formed in large amounts under certain conditions, and it can be purified easily and with good yields (Henning *et al.*, 1962). The amino acid composition of the A protein has been determined and the N- and C-terminal peptides have been isolated (Carlton and Yanofsky, 1962, and unpublished). The results of these studies suggest that the A protein is a single polypeptide chain containing approximately 280 amino acid residues.

Mutant strains which are unable to synthesize a functional A protein can be readily isolated following treatment of the wild-type strain with any one of a variety of mutagenic agents. These mutants are characterized as tryptophan auxotrophs that respond to indole (due to the presence of the normal B protein) and accumulate indoleglycerol.

Approximately 200 A-protein mutants have been isolated to date. These mutants were induced with ultraviolet light, 2-aminopurine, or ethylmethane sulfonate. All have mutational alterations in one region of the *E. coli* chromosome, which we have appropriately designated the A gene. The relative locations of the mutational alterations in many of these mutants has been determined, using 2-point and 3-point genetic tests (Yanofsky *et al.*, 1961, and Yanofsky, unpublished). Genetic studies have been carried out with the transducing phage Plkc (Lennox, 1955). The gene which determines the structure of the B protein of tryptophan synthetase is immediately adjacent to the A gene on the *E. coli* chromosome (Yanofsky, 1960). No revertable mutations affect both proteins.

Comparision of Gene Alterations and Protein Alterations

Enzymatic and immunological studies with extracts of the various A-protein mutants have shown that the mutants fall into two groups (Yanofsky and Crawford, 1959). Members of one group form a protein which resembles the

A protein in its immunological and enzymatic properties, while members of the second group lack any protein which can be recognized as A protein on the basis of these or other properties. There are several possible explanations for the absence of a detectable A protein in some of these strains: the protein could be formed but is extremely labile, the protein could have a tertiary structure considerably different from that of the wild-type protein, or the mutation could result in a nonsense coding unit (Crick *et al.*, 1957), a coding unit that does not specify an amino acid, and a complete polypeptide chain would not be formed.

The mutational alterations in the mutants that form altered A protein are located in four regions in the A gene (Fig. 1). Within each region there are many mutants with alterations at the same genetic site. Some of these strains would be expected to have the same mutational alteration (nucleotide substitution), while others could have different changes at the same site. In view of these considerations the properties of the proteins formed by these strains were examined for differences (Maling and Yanofsky, 1961; Henning and Yanofsky, 1963). The results of these studies are summarized in Fig. 1 where it can be seen that many distinguishable protein types are observed among the altered A proteins. It is also apparent that mutants with identical reversion patterns and map locations produce the same type of altered proteins, while other mutants altered at the same site form proteins with different properties (Fig. 1). These observations suggest that many of the mutants represent repeat identical mutations, presumably involving the same nucleotide substitution. However, two mutants that are indistinguishable by genetic and other criteria can only be established as identical by amino acid analyses of their altered peptides.

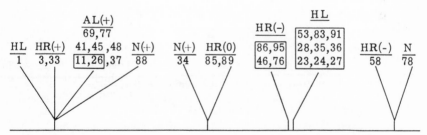

Fig. 1. Relative map location of the altered sites in the A mutants that form altered A proteins and the properties of the altered proteins. The mutants enclosed in the same box have been shown to have identical reversion patterns (Allen and Yanofsky, 1963). HL = heat-labile A protein, HR = heat-resistant A protein, AL = acid-labile A protein, N = normal stability to heat and acid; (+) (−) and (0) indicate electrophoretic migration relative to the wild-type protein.

Quantitative effects on enzyme formation are also observed. Several of the mutants that form altered A proteins consistently form lower levels of A protein than other mutants (Table II). Furthermore, the B protein level is also low in such strains. It is apparent from these findings that a mutational event may have two effects—it may lead to an alteration of the primary structure of a protein and it may also affect the ability of the organism to synthesize this protein and other proteins in the same pathway. Similar observations have been noted with other systems (Jacob and Monod, 1961; Lee and Englesberg, 1962); however, a satisfactory explanation of these findings has not been offered.

The various mutationally altered proteins have also been examined in fingerprinting studies (Ingram, 1961) with the intention of determining the amino acid change or changes associated with each mutational alteration. Using the fingerprinting approach, it was found that the A proteins of mutants in the A-23 group have a fingerprint pattern difference from the wild-type protein (Helinski and Yanofsky, 1962). This difference was shown to be the result of a substitution of a single amino acid, arginine, for glycine, at one position in one of the peptides of the A protein (Table III) (Helinski and Yanofsky, 1962). This amino acid change is apparently responsible for the loss of enzymatic activity of this protein and also for its characteristic heat lability.

TABLE II

THE LEVELS OF A AND B PROTEINS IN VARIOUS STRAINS[a]

Strain	A protein	B protein	Ratio A/B
Wild type	2.9	2.7	1.1
A-1	3.2	6.3	0.51
A-88	5.8	7.1	0.82
A-23	23	16	1.4
A-11 and most other A mutants	30–60	20–40	1–2

[a] The mutants were grown under identical conditions of derepression.

TABLE III

CORRESPONDING CHYMOTRYPTIC PEPTIDES FROM THE WILD-TYPE A PROTEIN AND THE A PROTEIN OF MUTANT A-23[a]

	1	2	3–5	6	7	8	9
Wild type	Asp-Ala-(Ala,Pro$_2$)-Leu-GluNH$_2$-Gly-Phe						
A-23	Asp-Ala-(Ala,Pro$_2$)-Leu-GluNH$_2$-Arg-Phe						

[a] Partial sequence based on unpublished data of Carlton and Yanofsky.

On the simplified map in Fig. 1 it can be seen that there is a second group of mutants, the A-46 group, which is genetically altered at a site very close to the altered site of the A-23 group. On the basis of the following considerations it was expected that the amino acid change in the A-46 protein would be at or near the position of the change in the A protein of A-23. The A gene is approximately 2.5 map units in length and the A protein contains 280 amino acids. The map length corresponding to a single amino acid is therefore 2.5/280 or approximately 0.01. The map distance between the A-23 and A-46 genetic sites is about 0.002, placing both changes within the map length corresponding to one amino acid. In view of these considerations the A-46 peptide was isolated that corresponds to the altered peptide of the A-23 protein. The analyses of this peptide showed that the A-46 peptide differed from the wild-type and A-23 peptides by the substitution of a glutamic acid residue at the position occupied by glycine in the wild-type peptide and arginine in the A-23 peptide (Henning and Yanofsky, 1962). Mutational changes that map very near each other therefore do lead to amino acid substitutions at the same position in the A protein. Several other mutants were examined that appeared to be indistinguishable from A-23 or from A-46 on the basis of other criteria and these were found to have the same amino acid substitutions detected in the A-23 and A-46 proteins.

Similar studies have been carried out with mutants with alterations that map at the other end of the A gene. Mutants A-1, A-3, A-11, and A-33 do not give wild-type recombinants when crossed to each other in all possible combinations (Maling and Yanofsky, 1961), suggesting that these mutants are altered at the same site in the A gene, and, consequently, in the same position in the A protein. The studies that have been performed with the proteins of these strains have not as yet led to the identification of the specific amino acid substitutions associated with the mutations in these strains. Nevertheless, they have shown that the three mutants that have been examined (A-3, A-11, and A-33) form proteins that have alteration in the same peptide of the A protein; and furthermore, in two of the three mutants the same wild-type amino acid appears to be replaced (Helinski and Yanofsky, 1962, and unpublished). Here again, the genetic data correlate with the position of amino acid substitutions in the protein, leading one to conclude with some confidence that there is a direct correspondence between gene structure and protein structure.

Reversion studies have also been carried out in an examination of the relationship between the genetic map and the position of amino acid changes in the protein. Reversion even when studied with point mutants includes several possible events (Table IV). Some mutants may result from additional nucleotide changes in the particular coding unit already altered in the mu-

TABLE IV

THE THREE CLASSES OF REVERTANTS RESULTING
FROM SINGLE NUCLEOTIDE SUBSTITUTIONS

Wild type	Mutant	Revertant	Reversion change
A	A	A	(1) Same nucleotide as in mutant
C--------→T--------→G			
C	C	C	
A	A	A	(2) Different nucleotide in same coding unit
C	C--------→A		
C--------→G	G		
A	A	A	(3) Nucleotide in different coding unit
C--------→T	T		
C	C	C	
⋮	⋮	⋮	
A	A	A	
A--------→G			
T	T	T	

tant, either at the same or a different nucleotide position. Others may be formed by a change in a different coding unit, but within the same gene (Table IV). In the latter case, termed second-site reversion, one would expect that the protein formed by the revertant would have two amino acid changes, one characteristic of the original mutant and a second due to the second-site reversion. Reverse mutations must be distinguished, of course, from suppressor mutations, which are mutations of other genes that reverse the effect of the original mutation. Suppressor mutations will be dealt with subsequently in this paper.

The most informative data on reversion of A mutants have been obtained from studies on the two mutants with known amino acid substitutions, strains A-23 and A-46. Both of these mutants revert to tryptophan-independence spontaneously, and some of these revertants result from changes in the *same* coding unit that was altered by the original mutation, while others are second-site reversions (Allen and Yanofsky, 1963; Helinski and Yanofsky, unpublished). In studies with mutant A-46 three classes of revertants have been distinguished. Members of two of these classes form A proteins that are considerably less active catalytically than the normal A protein (Yanofsky *et al.*, 1961). These strains do not synthesize tryptophan at the wild-type rate and thus grow considerably slower than the wild-type strain in the absence of tryptophan. We designate strains of this type "partial revertants." One of these partial revertant classes—the slower-growing class—results from a reversion at a second site, approximately one-tenth of the gene length away from the site of the original A-46 mutation (Helinski and Yanofsky, unpublished). A different partial revertant type appears to result from primary-site reversions. In addition to partial revertants, full revertants are obtained from both mutants. Full revertants are strains which are indistinguishable from wild-type in growth rate in a medium lacking exogenous tryptophan.

The A proteins of many of these partial and full revertant strains have been isolated and examined for amino acid changes. The protein of a second-site revertant of A-46 has been shown to retain the A-46 amino acid substitution; that is, glutamic acid for glycine in peptide TP3 (see Table V). However, in addition, a tyrosine in a different peptide, TP8, is replaced by cysteine (Helinski and Yanofsky, unpublished). This tyrosine to cysteine change apparently compensates for the glycine to glutamic substitution and the protein with both of these changes is slightly active. The second mutated site in this partial revertant strain has been separated from the A-46 site by appropriate crosses, and the strain with the second-site change alone is a mutant and forms a defective protein (Helinski and Yanofsky, unpublished). The protein of this strain has the tyrosine-cysteine change but has the wild-type amino acid glycine, in TP3 (Table V). Thus, in this case the substitution of the two amino acids, glutamic acid and cysteine, in the same protein, permits catalytic activity while either substitution by itself does not.

In genetic tests with the second partial revertant type derived from mutant A-46 it has not been possible to recover mutant A-46, suggesting that the reversion is at or near the site of the A-46 alteration. Analysis of the A protein formed by this strain showed that the glutamic acid of the A-46 protein

TABLE V

ANALYSIS OF SECOND-SITE REVERSION[a]

Strain	Amino acids in corresponding positions in relevant peptides		Protein	Location of genetic alteration
	TP[b]8	TP[b]3		
Wild type	Tyrosine	Glycine	Active	———————
A-46	Tyrosine	Glutamic acid	Inactive	—————|—
A-46PR8	Cysteine	Glutamic acid	Active	———|—|—
PR8	Cysteine	Glycine	Inactive	———————|—

[a] Based on unpublished studies of Helinski and Yanofsky.
[b] TP = tryptic peptide.

TABLE VI

<small>THE AMINO ACID SUBSTITUTIONS OBSERVED AT ONE POSITION IN THE A PROTEIN</small>

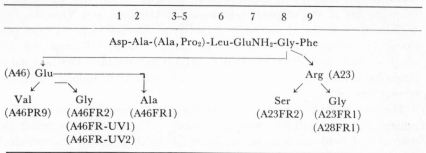

	1	2	3–5	6	7	8	9

Asp-Ala-(Ala, Pro₂)-Leu-GluNH₂-Gly-Phe

| (A46) Glu | | | | | | Arg (A23) |

Val	Gly	Ala			Ser	Gly
(A46PR9)	(A46FR2)	(A46FR1)			(A23FR2)	(A23FR1)
	(A46FR-UV1)					(A28FR1)
	(A46FR-UV2)					

has been replaced by valine (Table VI) (Henning and Yanofsky, 1963). Since the wild-type protein has glycine in this position it would appear that valine or glycine in this position in the A protein permits functional activity of this protein. However, the protein with valine is appreciably less active.

The proteins of two spontaneous full revertants and two ultraviolet-induced full revertants of A-46 have also been examined. The ultraviolet-induced reversions both restored glycine at position 8 (Table VI) (Yanofsky, unpublished). The A protein of one of the two spontaneous full revertants had an alanine instead of glutamic acid at position 8 (Table VI) (Allen and Yanofsky, 1963; Henning and Yanofsky, 1963), while the protein of the other spontaneous revertant had the wild-type amino acid, glycine. Thus besides valine, alanine can substitute for glycine at this position in protein, and the protein is functional.

Many types of full revertants and partial revertants are also obtained from mutant A-23 (Allen and Yanofsky, 1963). However, it has not, as yet, been possible to purify the A proteins of any of the partial revertants. The A proteins from three spontaneous full revertants of A-23 have been isolated and analyzed, two have the wild-type amino acid, glycine, instead of arginine in position 8 of the peptide, while the third has serine in this position (Table VI) (Henning and Yanofsky, 1963). Here again it is apparent that amino acids other than glycine can occupy position 8 in the peptide and the protein will be functional.

The Relationship of Amino Acid Substitutions to the Genetic Code

Inasmuch as these studies have detected a series of amino acid substitutions at the same position in the protein, each presumably representing a single mutational event, and possibly a single nucleotide change, it is of interest to compare these amino acid substitutions with the code letters that have been proposed for these amino acids by Nirenberg *et al.* (1963) and Ochoa (1963). This comparison is given in Table VII. It can be seen that each amino acid substitution is consistent with a single nucleotide change. In view of this consistency it seems reasonable to assign certain nucleotides to the same relative position in the different coding units. When this is done two of the letters in the coding units for glutamic, valine, glycine, and alanine, U and G, occupy the same relative positions (Table VII).

A test of the assignment of different relative positions to some of the nucleotides of these coding units can be performed in the form of a genetic recombination experiment. As outlined in Table VIII, a cross between a strain with the arginine protein and a strain with the glutamic protein, could give recombinants with the glycine protein if the C of arginine and the A of glutamic acid occupied different relative positions, and also recombinants with some other amino acid at the same position. Similarly, a cross between a valine partial revertant and the arginine mutant should give two types of recombinants, glycine recombinants and serine recombinants if the C of arginine and the U's of valine occupied different positions. These crosses

TABLE VII

COMPARISON OF AMINO ACID SUBSTITUTIONS WITH PROPOSED CODE LETTERS

(A23FR2)	Serine	UUC		UCG	Alanine	(A46FR1)
		↑			↑	
(A23)	Arginine	UGC ⇌ UGG ⇌		UAG	Glutamic	(A46)
					↓	
		Glycine	UUG		Valine	(A46PR9)
		(wild type)				

TABLE VIII

RECOMBINATION WITHIN CODING UNITS

Cross	RNA code letters[a]	Possible recombinant classes	Recombinants detected
Arginine × glutamic	UGC UAG	UGG (Glycine) UAC (?)	Glycine (2)
Arginine × valine	UGC UUG	UGG (Glycine) UUC (Serine)	Glycine (1) Serine (2)

[a] RNA code letters rather than DNA letters are used for simplicity.

were performed and it was found that all but one of the expected recombinant types were obtained (Table VIII) (Henning and Yanofsky, 1963). The fact that the reciprocal recombinant class, the presumed UAC type, was not detected in the arginine-glutamic cross may indicate that this coding sequence corresponds to an amino acid which leads to a nonfunctional protein. Only those recombinants with a functional A protein could be detected in these crosses. The recombination frequency between the various strains employed in these crosses is very low, of the order of 0.001–0.004%. Because of this low recombination value it is particularly important to exclude reversion of either donor or recipient as a possible source of the new amino acids that were detected in the presumed recombinants. Reversion studies performed with all the strains employed in these crosses have indicated that the reversion rates are too low to account for the appearance of the tryptophan-independent colonies recovered in these crosses.

The results of these recombination experiments confirm the relative order of the nucleotides that was assigned on the basis of the assumption that each mutational event involved a single nucleotide change. One unique feature of these recombinational analyses should be pointed out. In ordinary recombination studies a recombinant protein would be expected to have different combinations of amino acids already existing at specific positions in the parental proteins. However, in the crosses just discussed *new* amino acids are actually formed as a result of recombination *within* a coding unit.

The combined results of these studies demonstrate that the amino acid changes which have been detected can be related to different nucleotide alterations in a particular coding unit determining one amino acid position in the A protein. Furthermore, the changes detected are all consistent with the code letters assigned to the amino acids by studies of *in vitro* amino acid incorporation into polypeptide chains (Nirenberg *et al.*, 1963; Ochoa 1962). As yet no change has been detected in the presumed third letter of the coding unit—the common U.

Suppressor Mutations—Alterations of the Translation Mechanism

Suppressor mutations are, by definition, changes in one gene which reverse the phenotypic effects of a mutation in another gene. They, therefore, represent instances of gene interactions, and are of particular interest in relation to the question of whether the primary structure of a protein can be affected by more than one gene. The concept which is basic to the work that I have discussed so far is that the primary structure of a specific polypeptide or protein is determined by the nucleotide sequence of one gene. Inasmuch as suppressor mutations occurring in a second gene reverse the effects of some mutations, it is appropriate to ask whether this reversal is the result of alterations in protein primary structure.

Since suppression is defined on the basis of phenotypic change, it would be expected that suppressors could act in many ways. Studies of several cases of suppression have in fact revealed a variety of mechanisms of phenotypic reversal (see Fincham, 1960; Yanofsky and St. Lawrence, 1960). In studies on the effects of suppressor mutations on specific enzyme systems, we observed that suppressor mutations lead to the formation of enzymatically active protein when they occur in some strains which are incapable of forming enzymatically active protein (Yanofsky and Bonner, 1955; Yanofsky and Crawford, 1959). This result was obtained with mutants that formed an altered, detectable protein (Suskind *et al.*, 1955; Yanofsky and Crawford, 1959), and with strains that lacked detectable protein (Yanofsky and Crawford, 1959) (Table IX). Thus suppressor mutations affect both major classes of mutant types. Most of the suppressor genes that we have studied have been found to be extremely specific, affecting only the mutants in which

TABLE IX

THE ENZYMATIC ACTIVITY OF THE A AND B PROTEINS
FROM VARIOUS SUPPRESSED MUTANTS[a], [b]

Strain	$\dfrac{\text{InGP} \rightarrow \text{Tryp}^c}{\text{In} \quad \rightarrow \text{Tryp}}$	$\dfrac{\text{InGP} \rightarrow \text{Tryp}^c}{\text{InGP} \rightarrow \text{In}}$
Wild type	1	1
A2[d]	0	—
A2su	1	—
A3	0	—
A3su	0.01	—
A11	0	—
A11su	0.02	—
A36	0	—
A36su	0.1	—
B1	—	0
B1su	—	0.007
B4[d]	—	0
B4su	—	2.6

[a] All altered A proteins are active in the In → Tryp reaction and all altered B proteins are active in the InGP → In reaction when the normal second component is also present. The InGP → Tryp activity is presumably a measure of the wild-type-like protein in suppressed mutant extracts.

[b] In = indole; InGP = indole-3-glycerol phosphate; Tryp = tryptophan.

[c] Wild-type ratio arbitrarily set at 1.

[d] Does not form a detectable altered protein.

they were isolated, or mutants that are considered identical on the basis of other criteria. This finding in itself is suggestive that this type of suppressor mutation may be specific for particular amino acid changes in a protein. Other mechanisms of suppressor gene action, such as the activation of an alternate pathway would be expected to lead to the nonspecific suppression of many mutants with alterations in the same gene.

In an effort to determine whether suppressor mutations affect primary structure, we have examined the A proteins of three suppressed mutants. These three suppressed mutants were derived from strains which formed altered A proteins. These studies have shown that each of the three suppressed mutants forms two types of A protein, while the corresponding unsuppressed mutants form only one (Brody and Yanofsky, 1963). Of the two types of A protein formed by each suppressed mutant, one type is indistinguishable from the altered A protein of the original mutant while the other resembles the wild-type A protein, both enzymatically and in its physical properties (Brody and Yanofsky, 1963). With extracts of two of the suppressed mutants an attempt was made to separate the two A proteins by chromatography on DEAE-cellulose, and in both cases separation was achieved (Crawford and Yanofsky, 1959; Brody and Yanofsky, 1963). This observation suggests even more strongly that there are two types of A protein in the suppressed mutants. The amount of the second, wild-type-like protein in the suppressed mutants is very low, generally below 10% of that of the total A protein formed by the suppressed mutant (Table II). However, in one case an attempt is presently being made to isolate enough of the wild-type-like protein to examine its primary structure, especially in the region of the A protein where the amino acid replacement occurred in the original mutant. Preliminary results have been obtained in these studies which suggest that the wild-type-like protein that is formed has a primary structure which differs from the primary structure of the A protein of the original mutant (Brody and Yanofsky, 1963). These results were obtained in fingerprinting studies of the wild-type-like protein. The peptide pattern of the wild-type-like protein differs from that of the mutant protein by a single peptide and more closely resembles that of the wild-type protein (Brody and Yanofsky, 1963). This finding, however, does not establish that the primary structure of the wild-type-like protein is identical to that of the wild-type A protein. However, it can be concluded that a primary structure change has occurred in a fraction of the A protein molecules formed by the suppressed mutant, and that the resultant protein is indistinguishable from the wild-type protein in all of the characteristics that have been examined. Therefore, although we cannot specify the change, it is clear that this type of suppressor mutation can alter protein primary structure.

A reasonable possibility by which mutations in suppressor genes could lead to amino acid replacements in proteins is by changes in the transfer ribonucleic acid (RNA) system. This mechanism has also been considered by M. Lieb and L. Herzenberg (personal communication) and by Benzer and Champe (1961). Suppressor mutations could alter either the composition of transfer RNA molecules or the amino acid activating enzymes so that incorrect amino acids are occasionally attached to a particular transfer RNA molecule. Or, the transfer RNA molecules could occasionally pair incorrectly with the RNA template. This could lead to an amino acid incorporation mistake and a protein with the primary structure of the wild-type protein. A less likely alternative for the presence of both mutant and wild-type proteins is that suppressor genes alter some RNA template molecules at specific locations. In an attempt to examine the first possibility, two lines of investigation are being carried out (Brody and Yanofsky, unpublished). The amino acid composition is being determined on the total protein of strains with or without specific suppressor mutations. If suppressor mutations do alter amino acid incorporation specificity, mistakes should be distributed at random throughout all the proteins of Escherichia coli, and, thus, the total amino acid composition should be affected. Of the suppressed mutants that have been examined one does appear to have amino acid composition alterations (Brody and Yanofsky, 1963). Several amino acids are high and others are low relative to the protein of strains without the mutated suppressor gene. The two amino acids that are affected to the greatest extent are histidine and tyrosine; the histidine of the suppressed strain is low by about 8%, while the tyrosine is high by about the same amount. Transduction tests carried out with this suppressed mutant have shown that the deviation with respect to histidine and tyrosine content is associated with the suppressor gene (Brody and Yanofsky, 1963), although this certainly does not establish that the suppressor gene itself is responsible for the histidine-tyrosine change or the other amino acid composition changes. The finding of amino acid composition changes would appear to support the view that suppressor mutations can affect amino acid incorporation specificity; however, the composition changes detected in the suppressed mutant do not account for the results of fingerprinting studies with the same strain. The A protein of the original mutant has a glycine to arginine change and the composition of neither of these amino acids is affected by the mutated suppressor gene. Fingerprinting studies with the wild-type-like protein of the suppressed mutant suggest that the arginine is replaced by some other amino acid (Brody and Yanofsky, 1963). This amino acid need not be glycine since it is certainly clear from our other studies that there are many amino acids that could replace arginine at the appropriate position in the protein, and the protein would be functional. Thus, although a protein resembling the wild-type protein is restored by suppressor mutations, it need not have the exact amino acid sequence as the wild-type protein. Nevertheless, it would have a primary structure change, which is the important finding when mechanisms are considered.

The second approach we are employing is to examine the specificity of the transfer RNA systems in the mutants and in the suppressed mutants. Although these studies are under way, the critical experiments have not as yet been performed.

It is clear from our studies with suppressor genes that the primary structure of the A protein can be altered by mutations outside the A gene. Regardless of the mechanism by which this primary structure change is accomplished, it is obvious that organisms have the ability to translate a nucleotide sequence into two or more slightly different proteins.

Acknowledgments

These investigations were supported by grants from the National Science Foundation and the United States Public Health Service.

References

Allen, M. K., and Yanofsky, C. (1963). In preparation.

Brody, S., and Yanofsky, C. (1963). In preparation.

Benzer, S., and Champe, S. P. (1962). Ambivalent r II mutants of phage T4. *Proc. Natl. Acad. Sci. U.S.* **47,** 1025–1038.

Carlton, B. C., and Yanofsky, C. (1962). The amino-terminal sequence of the A protein of tryptophan synthetase of *Escherichia coli. J. Biol. Cnem.* **237,** 1531–1534.

Crawford, I. P. (1962). An investigation of the B components of *Escherichia coli* tryptophan synthetase. *Bacteriol. Proc. (Soc. Am. Bacteriologists).* p. 119.

Crawford, I. P., and Yanofsky, C. (1958). On the separation of the tryptophan synthetase of *Escherichia coli* into two protein components. *Proc. Natl. Acad. Sci. U.S.* **44,** 1161–1170.

Crawford, I. P., and Yanofsky, C. (1959). The formation of a new enzymatically active protein as a result of suppression. *Proc. Natl. Acad. Sci. U.S.* **45,** 1280–1288.

Crick, F. H. C., Griffith, J. C., and Orgel, L. E. (1957). Codes without commas. *Proc. Natl. Acad. Sci. U.S.* **43,** 416–421.

Fincham, J. R. S. (1960). Genetically controlled differences in enzyme activity. *Advances in Enzymol.* **22,** 1–43.

Helinski, D. R., and Yanofsky, C. (1962). The correspondence between genetic data and the position of amino acid alteration in a protein. *Proc. Natl. Acad. Sci. U.S.* **48,** 173–183.

Henning, U., and Yanofsky, C. (1962). An alteration in the primary structure of a protein predicted on the basis of genetic recombination data. *Proc. Natl. Acad. Sci. U.S.* **48,** 183–190.

Henning, U., and Yanofsky, C. (1963). In preparation.

HENNING, U., HELINSKI, D. R., CHAO, F. C., AND YANOFSKY, C. (1962). The A protein of the tryptophan synthetase of *Escherichia coli*. *J. Biol. Chem.* **237**, 1523–1530.

INGRAM, V. M. (1961). "Hemoglobin and Its Abnormalities." Thomas, Springfield, Illinois.

JACOB, F., AND MONOD, J. (1961). On the regulation of gene activity. *Cold Spring Harbor Symposia Quant. Biol.* **26**, 193–211.

LEE, N., AND ENGLESBERG, E. (1962). Dual effects of structural genes in *Escherichia coli*. *Proc. Natl. Acad. Sci. U.S.* **48**, 335–346.

LENNOX, E. S. (1955). Transduction of linked genetic characters of the host by bacteriophage P1. *Virology* **1**, 190–206.

MALING, B., AND YANOFSKY, C. (1961). The properties of altered proteins from mutants bearing one or two lesions in the same gene. *Proc. Natl. Acad. Sci. U.S.* **47**, 511–566.

NIREMBERG, M. W., MATTHAEI, J. H., JONES, O. W., MARTIN, R. G., AND BARONDES, S. H. (1963). Approximation of genetic code via cell-free protein synthesis directed by Template RNA. *Federation Proc.* **22**, 55–61.

OCHOA, S. (1963). Synthetic polynucleotides and the genetic code. *Federation Proc.* **22**, 62–74.

SUSKIND, S. R., YANOFSKY, C., AND BONNER, D. M. (1955). Allelic strains of *Neurospora* lacking tryptophan synthetase: A preliminary immunochemical characterization. *Proc. Natl. Acad. Sci. U.S.* **41**, 577–582.

YANOFSKY, C. (1960). The tryptophan synthetase system. *Bacteriol. Revs.* **24**, 221–245.

YANOFSKY, C., AND BONNER, D. M. (1955). Gene interaction in tryptophan synthetase formation. *Genetics* **40**, 761–769.

YANOFSKY, C., AND CRAWFORD, I. P. (1959). The effects of deletions, point mutations, reversions and suppressor mutations on the two components of the tryptophan synthetase of *Escherichia coli*. *Proc. Natl. Acad. Sci. U.S.* **45**, 1016–1026.

YANOFSKY, C., AND ST. LAWRENCE, P. (1960). Gene action. *Ann. Rev. Microbiol.* **14**, 311–

YANOFSKY, C., HELINSKI, D. R., AND MALING, B. D. (1961). The effects of mutation on the composition and properties of the A protein of *Escherichia coli* tryptophan synthetase. *Cold Spring Harbor Symposia Quant. Biol.* **26**, 11–23.

Genetic Repression, Allosteric Inhibition, and Cellular Differentiation

FRANÇOIS JACOB AND JACQUES MONOD

Services de Génétique microbienne et de Biochimie cellulaire, Institut Pasteur, Paris

Introduction

It is generally recognized today that the characteristics of an individual, its development and functioning, are written in a codescript along its chromosomes. Recent work in biochemistry and in genetics indicates that, in every cell, a given nucleotide sequence determines, via a corresponding amino acid sequence, a particular function different from those determined by other nucleotide sequences, and this is so for thousands of different functions.

This structural concept of gene action accounts for the multiplicity and for the phylogenetic stability of macromolecular structures. It does not account for the physiological coordination of chemical activity, i.e., of synthesis and activity of macromolecules, which is a fundamental requirement for the existence and survival of the cell as well as of the multicellular organism. The complex and precise chemical network of information transfer upon which the development and physiological functioning of organisms must rest, implies the existence of precise regulatory systems at the level of both the organism and the cell.

For obvious technical reasons, the analysis of intracellular mechanisms of regulation have, so far, largely been restricted to microorganisms, where two basic devices have now been recognized. One regulates the *activity* of certain enzymes, and thereby insures a rapid and sensitive control of metabolic pathways, by the process usually called *feedback inhibition*. The other regulates the biosynthesis of enzymes by the process of *genetic repression*. The question which we wish to consider is: to what extent do the basic mechanisms found to operate in bacteria also apply to cells of higher organisms, whose functions are performed under very different and far more complex conditions? More particularly, may the concepts derived from the study of regulation in bacteria be of some value in the interpretation and analysis of cellular differentiation?

Cellular differentiation controls the time of emergence, the shape, the number, and the functions of cells, their organization into tissues and specialized organs. As a result of its complexity, differentiation has been defined in a variety of ways. In this discussion, we shall deliberately oversimplify the problem and restrict our argument to one aspect of this problem. We shall consider that two cells are differentiated with respect to each other if, while they harbor the same genome, the pattern of proteins which they synthesize is different. This definition emphasizes one of the main difficulties in the interpretation of cellular differentiation. The cells of an organism have evolved from a common cell. In all likelihood, they possess identical chromosomal sets and, on the basis of the structural gene theory, they would be expected to synthesize the same proteins and, therefore, to perform identical functions. Yet in the course of development, different types of cells which perform different functions progressively emerge and this diversity is clonally transmitted in a stable way. Whether or not differentiation is irreversible has long been debated. The answer, however, still remains ambiguous: while reversibility may directly be demonstrated in some instances, absolute irreversibility is hardly a meaningful concept, since it evades any strict operational definition and experimental test. In fact, although genetic analysis of tissue cells remains to be done, most of the known facts support the view that the genetic potentialities of differentiated cells have not been fundamentally altered, lost, or distributed. One of the strongest arguments comes from the study of plants, where undoubtedly cells are morphologically, physiologically, and biochemically differentiated although any single cell, even if differentiated, is still capable of giving rise to a complete organism (see Braun, 1961). Differentiation is probably not irreversible, but is certainly stable. What must be explained are the systems preventing any cell from expressing all its genetic potentialities and the stable transmission of the signals involved in the sorting out of different functions.

In this paper, we wish to describe the basic systems of regulation observed in bacteria and to discuss their use as models for interpreting the emergence and maintenance of differentiated cellular lineages within a genetically homogeneous population.

Regulation of Enzymatic Activity: Allosteric Effects

In many enzyme systems, enzymatic *activity* is regulated by metabolites unrelated, structurally, to the substrates of the regulated enzyme. By the definitions which we adopted above, such effects do not involve any differentiation, since they do not alter the pattern of protein synthesis. These phenomena, however, do belong in this discussion not only because of their physiological importance, but also as models of a general class of interactions, designated as "*allosteric* effects" (Monod and Jacob, 1961), which, we believe, may be of fundamental importance for the interpretation of biological regulation in general, including differentiation.

The study of these effects, in bacteria, stems from the discovery by Novick and Szilard (1954) that the addition of tryptophan to the growth medium of *Escherichia coli* results in immediate cessation of tryptophan synthesis by the cells. Their work led Novick and Szilard to the conclusion that tryptophan acted as an inhibitor of an *early* enzyme of the tryptophan-synthesizing-pathway. The pioneer work of Umbarger (1956, 1961) confirmed and extended by several others (see references in Frisch, 1961) showed that similar "feedback inhibition" of an enzyme by a distant product of its activity occurs not only in the tryptophan system, but actually in most if not all pathways leading to the synthesis of essential metabolites. Besides their obvious physiological significance, these effects propose some very interesting problems in enzymology. In most of the cases which have been studied so far, the inhibition is "competitive" in the sense that it depends upon the *relative* concentration of inhibitor and substrate. Now, as is well known, competitive inhibitors of enzymes are in general structurally related to the substrate: they are isosteric (or partially isosteric) *analogs* of the substrate. The regulatory enzymes (i.e., those responsible for the Novick-Szilard effect) appear to violate this rule, since they are inhibited by substances which are sterically unrelated to their substrate (allosteric). Many types of more or less nonspecific inhibition of enzymes are known, of course, and allosteric inhibition of regulatory enzymes would not be so remarkable, if it were not for the extreme specificity of the effect. Threonine deaminase, for instance, is inhibited powerfully and competitively by isoleucine (a distant product of threonine deamination), but not by any other naturally occurring amino acid (Umbarger, 1956). The inhibition therefore cannot be due to the structural features common to isoleucine and threonine, since these are shared by many other amino acids. Similarly, aspartic transcarbamylase (ATCase) (Yates and Pardee, 1956) is inhibited by cytidine triphosphate (CTP), which is hardly an analog of aspartic acid, although part of the pyrimidine ring does derive (through many enzymatic steps) from the substrate of ATCase.

Another, highly significant criterion of the specificity of these effects is to be seen in the fact that in all of the regulated pathways studied so far, only one enzyme, as a rule the first one in the pathway concerned (Stadtman *et al.*, 1961), is sensitive to allosteric inhibition. Conversely, none of the intermediates formed in the pathway are active as inhibitors, this function being, in all cases, fulfilled by the final product. These observations alone would suffice to indicate that sensitivity to allosteric inhibitors results from a highly specialized and exceptional "construction" of the enzyme protein molecule itself.

Studies of the kinetics of the reactions catalyzed by allosteric enzymes have brought direct evidence on this point. In most of the cases which have been studied so far, the kinetics of action of substrate or inhibitor, or both, turn out to be quite different from the classical Henri-Michaelis kinetics usually observed in the case of classical enzymes. Since we could not in the present paper go into a description and analysis of these observations, we may perhaps summarize them by indicating that the reaction catalyzed by allosteric enzymes very often, if not as a rule, obeys multimolecular rather than monomolecular relations with respect to both substrate and inhibitor. As shown by the work of Changeux (1961) and also apparent from the results of Gerhart and Pardee (1962), these kinetics cannot be accounted for by the assumption, adequate in the case of normal enzymes, that the binding of substrate and inhibitor occurs at the same site of the enzyme surface and are mutually exclusive. As pointed out by Changeux, in particular, one is led to the conclusion that substrate and inhibitor actually bind at two (or at least two) different sites, and may actually be simultaneously bound by the enzyme. This conclusion is supported by the remarkable observation (Changeux, 1961; Gerhart and Pardee, 1962; Martin, 1962) that various treatments known to be capable of partially inactivating or denaturing enzymes in general, may result, in the case of allosteric enzymes, in desensitization (loss of sensitivity to inhibitor) without loss of activity toward substrate. Moreover, desensitization of the enzyme is attended by "normalization" of its kinetics with respect to substrate. Finally, in at least two instances, it has been definitely shown that such desensitization also results in alterations of the sedimentation velocity of the proteins (Gerhart and Pardee, 1962; Martin, 1962). The sum of these observations very strongly suggests that the action of allosteric inhibitors is not due to a *direct* interference, by steric hindrance, with the binding of substrate, but rather to an induced alteration of the shape or structure of the enzyme protein, resulting in misfit or reduced fit of the substrate at the active site.

If this is indeed true, one might also expect that allosteric effects might also operate positively, i.e., by increasing the fit of the substrate at the active site. Indeed Changeux (1962) has found that while isoleucine is an inhibitor, valine is an activator of threonine deaminase and Gerhart and Pardee (1962) have observed that while ATCase is inhibited by CTP, adenosine triphosphate (ATP) is a potent activator of the reaction.

In the discussion so far we have mostly considered observations made with bacterial enzymes, where the physiological, actually the nutritional, significance of the regulation is obvious. There are now many reports in the literature indicating that similar mechanisms operate with certain enzymes extracted from tissues of higher organisms. One of the best studied and most striking cases is glycogen synthetase discovered by Leloir and Cardini (1957) and studied by Algranati and Cabib (1962) and Traut (1962). This enzyme which synthesizes glycogen from uridine diphosphoglucose (UDPG) is strongly activated by glucose-1-phosphate and also inhibited by ATP. While glucose-1-phosphate is of course sterically identical with the glucose moiety of UDPG, it is obvious that its activity is not due to binding at the active site of the enzyme and the careful study of Traut indicates that its effect and also the effect of ATP are due to an induced alteration of the structure of the enzyme protein itself.

Of particular interest to the present subject are those cases where the allosteric agent is a hormone. The best known example is found in the beautiful work of Tomkins and Yielding (1961) on glutamic dehydrogenase. As is well known, Tomkins discovered that, in presence of certain steroids, this protein largely loses its activity toward glutamic acid while acquiring activity toward alanine. Furthermore, the transition is accompanied by depolymerization of the enzyme and these effects are antagonized or reverted by adenosine diphosphate (ADP), as well as by certain amino acids. Several other cases involving alterations of enzyme activity by steroids have been reported, and the multiple effects of cyclic adenosine monophosphate (AMP) may also be recalled here (Rall and Sutherland, 1961), although the mechanism of these effects has not yet been clarified.

One might ask, at this point, whether allosteric agents are clearly distinct from coenzymes. The brief summary given above suffices to indicate that coenzyme action could hardly be described in similar terms; not only because, in contrast to allosteric agents, coenzymes are active with many different enzymes catalyzing similar reactions, but also and mainly because the true coenzymes are known to participate directly in the enzyme reaction by forming an intermediate covalent compound with the substrate or part of the substrate molecule: coenzymes actually behave as second substrates. Although the study of allosteric effects is quite recent, it already seems clear that typical allosteric effectors do not participate in the reaction itself and there is no indication, in the best studied instances, that they undergo any covalent reaction (Traut, 1962).

This being said, it should be recalled here that, according to Koshland (1959), enzyme-substrate interaction, in general, may involve "induced fit", i.e., mutual effects of substrate and enzyme on the molecular configuration of each. If Koshland's theory is correct, allosteric effects might be considered as an extension and specialization of a basic mechanism common to most or all biologically active proteins.

While the physiological interpretation of most allosteric effects observed with enzymes of higher organisms is not simple and immediate, as with bacterial systems, it can hardly be doubted that these mechanisms do have a regulatory function. And although the observations which we have briefly summarized above concern very different systems operating in widely different organisms, they all would seem to show, rather strikingly, some common features which define them as belonging to the same general class: namely, enzyme proteins whose specific activity or affinity toward their substrate is selectively increased or decreased by agents which do not act by virtue of either being analogs of the substrate or actual intermediary participants in the reaction, but by binding with the enzyme protein at a site distinct from the active site, such binding resulting in alterations of the molecular structure of the protein.

In any case, and whatever their precise mechanism may be, the most important point about allosteric effects, from a biological point of view, is the absence of any direct chemical relationship between substrate and allosteric inhibitor or activator: all the evidence points to the conclusion that the particular, specific effect of a given allosteric agent is due exclusively to a highly specialized construction of the competent protein itself. This means that proteins subject to allosteric effects are to be considered as pure products of selection for adequate regulatory interactions.

The interpretative power of this concept is evidently very great. Indeed it is so great that it should be used with some caution. There is at present no direct proof that allosteric effects are at the basis of phenomena of differentiation as defined in the introduction. We shall return to this problem after having discussed genetic control of protein synthesis and genetic regulation.

Regulation of Protein Synthesis in Bacteria

Bacteria of identical genotype grown in different media do not exhibit the same enzyme patterns (see references in Frisch, 1961). For instance, when *Escherichia coli* is grown in the absence of tryptophan, the enzymes of the pathway involved in tryptophan production are actively synthesized. As soon as tryptophan is added to the medium, the enzymes cease to be synthesized. This effect of tryptophan, which is called *repression* of enzyme synthesis is extremely specific. In the same way, when *E. coli* is grown in the absence of a galactoside, only traces of the enzyme β-galactosidase are

formed by the cell. As soon as a galactoside is added, the rate of synthesis of
this enzyme increases by about 10,000-fold. Removal of the inducer results
in an almost instantaneous arrest of enzyme synthesis. This effect, called
induction, is also very specific. Finally, the production of phage by lysogenic
bacteria provides another example of change in the expression of genetic
potentialities. In lysogenic bacteria, the genetic material of the phage is
carried in the prophage state, the viral deoxyribonucleic acid (DNA) repli-
cates at the pace of the bacterial chromosome, and viral proteins are not
formed. Only as a result of a change in cellular conditions, either spontaneously or induced by various agents, such as ultraviolet light, X-rays, or
various chemicals, do the viral genes express their potentialities and the
bacteria produce viral particles.

The three systems used as examples are at first sight widely different. Yet
the results of genetic analysis and biochemical characterization of mutations
affecting regulation in these three systems are so closely similar that they
point to a common basic mechanism operating in all three systems.

We shall now describe the general model for the control of protein synthesis in bacteria which can be constructed on the basis of this analysis. This
model, diagrammatically represented in Fig. 1, involves the following points:

1. The structure of a protein (or polypeptide chain) is determined by a
particular deoxynucleotide segment, or *structural gene*. The primary product of
the structural gene is a short-lived ribonucleic acid (RNA) copy of the gene,
or *messenger RNA*, which brings structural information to cytoplasmic protein-forming centers. Once completed, messenger RNA is detached from
DNA and associates in the cytoplasm with pre-existing, nonspecialized

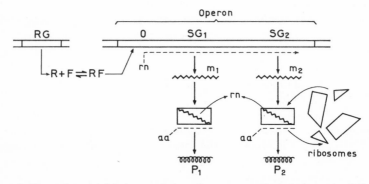

FIG. 1. General model for the regulation of enzyme synthesis in bacteria. RG: regulator gene; R: repressor which associates with effector F (inducing or repressing metabolite); O: operator; SG_1, SG_2: structural genes; rn: ribonucleotides; m_1, m_2: messengers
made by SG_1 and SG_2; aa: amino acids; P_1, P_2: proteins made by ribosomes associated
with m_1 and m_2.

ribosomal particles. The transcription of the genetic information from nucleotide sequence into amino acid sequence takes place on the ribosome and the messenger is rapidly destroyed in the process, after few copies of the polypeptide chains have been produced. Once completed, the polypeptide folds and is detached from the ribosomal particle which is set free for a new transcription cycle involving the same, or any other specific, messenger.

2. The synthesis of messenger RNA is assumed to be a sequential and oriented process which can be initiated only at certain regions, or *operators*, of the DNA strands. In some instances, a single operator may control the transcription of several adjacent structural genes into messenger RNA. The unit of primary transcription thus coordinated by a single operator constitutes an operon. An operon may contain one or several adjacent structural genes, depending on the system.

3. The rate of transcription of structural genes is negatively controlled by other, functionally distinct, determinants called *regulator genes*. A regulator gene forms a cytoplasmic product or *repressor*. The repressor formed by a given regulator gene has affinity for, and tends to associate reversibly with, a specific operator. This combination blocks the production of messenger RNA by the whole operon controlled by the operator and therefore prevents the synthesis of the protein governed by the structural genes of the operon.

4. A repressor has the property of reacting with certain small molecules, which we shall call effectors. The reactions are specific with respect to both the repressor (R) and the effector (F) and may be expressed as

$$R + F \rightleftharpoons RF$$

In certain systems, called inducible, only the R form of the repressor is active and blocks the transcription of the operon. The presence of the effector (called inducer) inactivates the repressor and therefore allows messenger synthesis to take place. In other systems (called repressible), only the combined RF form of the repressor is active. Synthesis of messenger RNA by the operon, allowed in the absence of the effector (or repressing metabolite), is therefore prevented in its presence.

The experimental arguments leading to the formulation of this model have been described in detail in several papers (Jacob and Monod, 1961a, b; Monod *et al.*, 1961). We shall restrict our discussion to the major features of this model (Fig. 1).

Negative Regulation Controlled by Specific Genetic Determinants

The first major feature of the model is that the synthesis of proteins in bacteria is controlled by two kinds of genes: those which determine the specific structure of the proteins and those which regulate (negatively) the rate of information transfer from structural genes to proteins. The existence of this double genetic determinism is proved by the study of mutations affecting protein synthesis. Some mutations result in an alteration of the structure of a protein without disturbing the rate of its synthesis; these mutations define the structural gene controlling this protein. Other mutations, located in another small region of the bacterial chromosome, affect the rate of synthesis of the protein, or the response to specific compounds, without altering the structure of the protein; these latter mutations define the regulator gene.

In the different systems studied, a variety of alleles for the regulator genes has been found. From their properties, summarized in Table I, the mode of action of regulator genes can be analyzed. Constitutive (or non-repressible) R^- mutations result in a loss of the regulating device: the corresponding enzyme is synthesized at maximal rate, irrespective of the presence of specific inducing (or repressing) metabolite. Diploid heterozygotes R^+/R^-, however, are normally inducible (or repressible), a result which characterizes the main properties of the system. The fact that a single regulator gene controls the expression of both chromosomes demonstrates the existence of a *cytoplasmic product of the regulator gene* (*the repressor*). Furthermore, the *repressor operates negatively*, i.e., inhibits protein synthesis, since the active system (R^+) prevents protein synthesis while the inactive system (R^-) allows protein synthesis at maximal rate. This conclusion is also supported by the properties of the R^t alleles, found both in the inducible system of β-galactosidase (Horiuchi *et al.*, 1961) and in phage (Sussman and Jacob, 1962): at low temperature, the repression is active and the systems operate under the same conditions as in the wild type (R^+). At high temperature, the repression systems are inactivated, an effect which results in a constitutive synthesis of β-galactosidase or in the production of phage. This type of mutation has rather dramatic consequences in phage. Lysogenic bacteria carrying such a mutant R^t prophage can easily grow at low temperature. When shifted to high temperature, however, all the bacteria lyse and release phage. The mutant prophage behaves as a thermosensitive lethal factor for the host.

These repression systems are able to react with specific metabolites present inside the cell or introduced from outside. That the metabolite reacts with the repressor is supported by the properties of R^s alleles. In the case of β-galactosidase, for instance, the repressor synthesized by the R^s allele cannot be antagonized by inducers at normal concentrations. They produce small amounts of enzyme only when the concentration of inducer is 100 to 1000 times greater than that required for maximal enzyme synthesis by the

TABLE I

ALLELES OF REGULATOR GENES FOR DIFFERENT SYSTEMS OF *Escherichia coli*

Allele	Product	Properties in the system of		
		Enzymes for lactose utilization	Enzymes of biosynthetic pathways	Temperate phage λ
R^+	Wild type repressor	Inducible by specific inducers (β-galactosides).	Repressible by the terminal product of the pathway.	Able to lysogenize. Lysogenic system induced to phage production by exposure to ultraviolet light and various chemicals, or thymine starvation.
R^-	Inactive repressor	Constitutive. R^+/R^- heterozygotes are inducible.	Non-repressible. R^+/R^- heterozygotes are repressible.	Unable to lysogenize alone. In mixed infection with R^+ produce R^+/R^- double lysogenics.
R^s	Super-repressed. Repressor non-antagonizable by inducer.	Unable to produce enzymes even in presence of inducer. R^s/R^+ or R^s/R^- heterozygotes do not produce enzymes.[a]		Able to lysogenize. Non-inducible by exposure to ultraviolet or thymine starvation. R^s/R^+ or R^s/R^- double lysogenics are non-inducible.
R^t	Repression system susceptible to temperature.	Inducible at low temperature. Constitutive at high temperature.		Able to lysogenize at low, but not at high, temperature. Lysogenics grown at low temperature produce phage when shifted to high temperature.
R^r	Reverted system	Partially constitutive. Repressed by β-galactosides.		

[a] Some R^s mutants for lactose utilization produce small amounts of enzymes in the presence of concentration of inducer 100 to 1000 times greater than that required for maximal production by the wild type.

wild type. The R^s mutation results in a decrease of affinity of the repressor for the inducers (Willson *et al.*, 1963). R^s mutants are therefore unable to synthesize β-galactosidase under physiological conditions. Diploid heterozygous R^s/R^+ are also unable to synthesize the enzyme, a striking result from the genetic point of view since an R^s mutation corresponds to a dominant loss of function. The interaction between the repressor (R) and the specific metabolite or effector (F) may be represented as a reversible association into a complex RF:

$$R + F \rightleftharpoons RF$$

In an inducible system, the repressor R, product of the regulator gene, is active and prevents enzyme synthesis. The complex RF is inactive and the enzyme is synthesized. In repressible systems, R is inactive and in the absence of the metabolite, the enzyme is produced. The complex RF is active and prevents enzyme synthesis.

This mechanism, assumed at first to account by a single type of interaction for both inducible and repressible systems, is now supported by the properties of another allele (R^r) obtained in β-galactosidase (Willson *et al.*, 1963). The R^r mutant is partially constitutive, but the addition of galactosides which are inducers in the wild type results in a decrease of enzyme production, i.e., in a repressing effect. A mutation in a regulator gene may, therefore, convert an inducible system into a repressible one, showing that similar elements and interaction operate in both types of system. The existence of such mutational changes has obvious implications for the evolution of regulatory systems.

Regulator genes may be visualized as transmitters of cytoplasmic chemical signals, the repressors, which act negatively on the transcription of structural genes into proteins and can be either inactivated (induction) or activated (repression) by specific metabolites. The repressor is clearly defined as the product of a regulator gene, which can exhibit different properties depending on the regulator allele. We shall return later to the problem of the nature of the repressor and of its interaction with small molecules.

Operators and Polygenic Operons

The second important feature of the model is that the bacterial chromosome contains units of transcriptive activity, or operons, coordinated by a genetic element or operator. The existence of operons and operators is evidenced by the properties of mutations which alter the rate of transcription of several adjacent structural genes located on the same chromosome. For instance, in the system for lactose utilization in *E. coli*, where two adjacent structural genes determine two distinct proteins, the operator is defined by the characters of a series of mutations clustered at one extremity of the operon, in the terminal part of the structural gene controlling β-galactosidase synthesis (see Fig. 2). One type of mutation, called operator-negative (O^0), results in a complete block of the transcription process along the two structural genes, and therefore in a loss of the ability to synthesize both enzymes. The other type, or operator-constitutive (O^c), results in a constitutive synthesis of both proteins; diploid heterozygotes O^+/O^c synthesize constitutively the two proteins determined by the structural genes located in position *cis* with respect to the O^c allele, i.e., in the same chromosome. These properties indicate that the operator does not act via a cytoplasmic product, but controls directly the transcription of the adjacent chromosomal segment, containing the two structural genes, as a single unit. The properties of the O^c allele indicate that the operator is the receiver of the controlling signals, i.e., the receptor of the repressor, and in all likelihood, the initiating point for the transcription of the whole operon. One must therefore expect the effects of inducing, or repressing, conditions to be quantitatively the same for different proteins controlled by different genes belonging to the same operon. This coordinated synthesis has been verified for the two proteins of the lactose system under a variety of conditions.

Similar observations have been made in various bacterial systems, where the genes controlling the different steps of a given biochemical pathway are known to be frequently clustered (Demerec and Hartman, 1959). The system for galactose utilization of *E. coli*, for instance, involves three enzymes, controlled by three adjacent genes and induced by galactose. Their grouping into a single operon is shown by the properties of an operator constitutive (O^c) mutation which maps at the extremity of the segment, on the galactoses epimerase side (Buttin, 1962) (see Fig. 2). It is worth noticing that in the two systems analyzed in *E. coli*, the operator lies on the same side of the operon with respect to the bacterial chromosome (see Fig. 2). If supported by further cases, this relationship would suggest a polarity of transcription alone the whole bacterial chromosome. In the system for arabinose utilization in *E. coli*, mutations of the O^0 type suggest that the three clustered structural genes determining the three enzymes of the system which are inducible by arabinose, are coordinated by a single operator (Lee and Englesberg, 1962).

In *Salmonella*, the genes controlling eight enzymes involved in the biosynthesis of histidine are clustered in a small segment of the bacterial chromosome, and the synthesis of all these enzymes has been shown to be quantitatively repressed by the terminal product of the pathway, histidine. Small

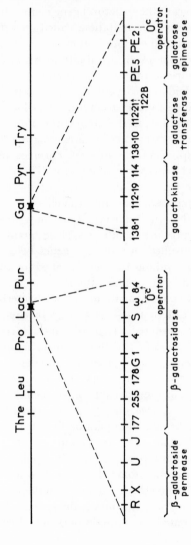

Fig. 2. Operons for the utilization of lactose and galactose in *Escherichia coli*.

deletions affecting a gene located at one extremity of the cluster (gene G, see Fig. 3), not only result in the incapacity to perform the first reaction of the pathway (controlled by the partially deleted gene G), but in a loss of the capacity to produce all the series of enzymes determined by intact genes (Ames and Garry, 1959; Ames *et al.*, 1960; Hartman *et al.*, 1960). The whole series of genes appear therefore to constitute an operon controlled by a single operator located at the extremity of the operon, on the side of gene G.

With this system, it has recently been possible to confirm a very distinctive prediction of the operon model. According to this model, the operator is the only receiver of controlling signals for the whole operon. Therefore, if a structural gene of the operon were physically disconnected from the operator, removed, and by some chromosomal rearrangement located somewhere else, the displaced structural gene would escape the control of its original operator and become insensitive to its normal system of regulation. From the strain, in which a partial deletion of gene G results in an inactivation of the whole series of genes, rare mutants have been isolated in which the series of intact genes have become again functional. Genetic analysis indicates that these mutants contain a double set of histidine genes: in addition to the normal set containing the original deletion and located in the normal region of the bacterial chromosome, they possess a duplicate set of the whole series of intact genes, which is not located in the usual histidine region of the chromosome, but in an unknown part of the cell genome (see Fig. 3). The two important points concerning these mutants are: (1) that the synthesis of the group of enzymes by the duplicated segment is not sensitive any more to repression by histidine, and (2) that the duplicated segment does not contain any detectable fragment of gene G (see Fig. 3) (Ames *et al.*, 1963). It is clear, therefore, that, while a deletion of the operator region results in a nonfunctioning of the whole operon, the activity of the intact genes may be restored by a physical disconnection from this operator. The fact that the transcription of the intact genes, thus detached, is not controlled any more by histidine demonstrates than the operator is indeed the only receiver of the specific regulating signals.

An operon considered as a unit of transcription may contain one or several structural genes, depending on the system. One repressor may act on a single operator or on several, the latter case being exemplified by the pathway of arginine biosynthesis of *E. coli*. In this system, the seven structural genes determining the synthesis of the seven known enzymes are distributed among five distinct regions of the bacterial chromosome. The synthesis of all enzymes is repressed by arginine and the available evidence suggests that a single

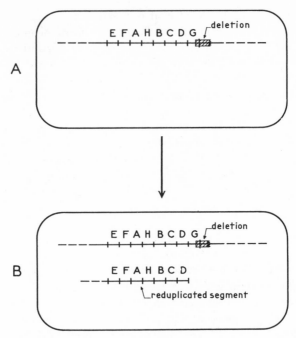

Fig. 3. The histidine operon in *Salmonella*. The eight structural genes controlling the enzymes of the pathway are clustered in a small segment of the bacterial chromosome. Deletions involving the extremity of the terminal gene G (which controls the first enzyme of the pathway, phosphoribosyl-ATP-pyrophosphorylase) result in the nonfunctioning of the seven other structural genes. From this strain (A), mutants (B) can be recovered (by the ability to grow on histidinol) in which the seven structural genes from E to D are functional. Cultures of these mutants segregate the original deletion (A) type. They carry a duplicate set of the seven genes, from E to D, which is not attached to the original histidine region. They do not possess any detectable part of the originally nondeleted portion of gene G. The synthesis of enzymes by the duplicated set of the seven genes is no longer sensitive to the repressing action of histidine. (From Ames *et al.*, 1963).

regulator gene, and therefore a single repressor, acts on the series of operators (Maas, 1961; Gorini *et al.*, 1961).

A similar situation is also found with the pyrimidine pathway in *E. coli*. The synthesis of the last four enzymes of the pathway is coordinately repressed by uracil, and the genes determining these enzymes are clustered in a small chromosomal segment, probably constituting an operon. The gene controlling the first enzyme is unlinked to the others, and the synthesis of this enzyme, while also repressible by uracil, is not quantitatively coordinated with that of the other enzymes. The whole system appears to be regulated by a single repressor acting on two operons, one constituted by four genes, and the other by one (Beckwith *et al.*, 1962).

Operons constitute the units of primary transcription, containing one or several genes. The operator may be visualized as the initiating point of the transcription process and the receiver of the controlling signals.

Messenger RNA and Regulation at the Genetic Level

The third, and most important, feature of the model is that regulation of protein synthesis operates at the genetic level, i.e., determines the rate of production of the primary gene product. This question is closely related to the problem of messenger RNA. The possible grouping of several structural genes into a single unit of activity indicates that regulation operates at a level where the structural information for several proteins is still associated in a single structure, which is probably the genetic material itself. This finding, together with the results of kinetic studies of the expression of a structural gene (Riley *et al.*, 1960), suggested that the transfer of structural information from genes to cytoplasmic protein-forming centers did not involve stable structures persisting in the cytoplasm, but rather metabolically unstable and rapidly renewed *messengers*. This conclusion resulted in a systematic search for an intermediate endowed with the proper characteristics, and several lines of evidence suggest that the so-called messenger RNA fulfills the prerequisites.

1. An RNA fraction with an exceptionally high rate of turnover exists, not only in phage-infected bacteria (Volkin and Astrachan, 1957), but also in normal bacteria (Gros *et al.*, 1961a) and in yeast (Yčas and Vincent, 1960). This fraction can reversibly adsorb to ribosomes. In contrast to other RNA fractions, its base composition approximates that of DNA (Gros *et al.*, 1961b) and it forms hybrids with homologous, but not with heterologous, DNA (Hall and Spiegelman, 1961), suggesting a direct relationship between the base sequence in DNA and in messenger RNA.

2. An enzymatic system able to synthesize RNA polynucleotides, using DNA as a primer and reproducing the DNA base ratio in its product, exists in bacteria (Weiss and Nakamoto, 1961; Stevens, 1960; Hurwitz *et al.*, 1961; Chamberlin and Berg, 1962). This product has the same characteristics as the "messenger RNA" fraction.

3. Ribosomes are not specific with respect to the proteins they synthesize. The same ribosomes are able to synthesize either bacterial or viral proteins, depending on whether the messenger RNA with which they are associated is bacterial or viral (Brenner *et al.*, 1961).

4. In reconstructed subcellular fractions, the presence of DNA appears

essential both for the synthesis of RNA, presumably messenger RNA, and for the incorporation of amino acids into peptide chains (Tissières *et al.*, 1960). Moreover, RNA polymerase has been shown to be an essential component in such systems (Wood and Berg, 1962). In the absence of DNA, incorporation of amino acids into peptide chains is stimulated by the messenger RNA fraction (Tissières and Hopkins, 1961).

The sum of these observations encourages the idea that it is indeed the primary gene product which has been identified as messenger RNA. The definite proof that messenger RNA does carry structural information to protein-forming centers is still lacking. However, the work with synthetic RNA polymers has clearly shown that ribosomes from various sources can be programmed by synthetic polymers (Nirenberg and Matthaei, 1961) and that different polymers lead to the formation of different products, a finding which constitutes a major step toward the solution of the coding problem (Lengyel *et al.*, 1961; Matthaei *et al.*, 1962).

As previously discussed, it is the result of genetic analysis and of studies of the kinetics of enzyme synthesis (following induction and/or gene transfer), suggesting that regulation of protein synthesis in bacteria occurred at the genetic level, which led to the hypothesis of an unstable messenger RNA as the carrier of genetic information from genes to protein-forming centers. Although the discovery of an RNA fraction which qualifies nicely as an unstable messenger certainly brings strong support to the hypothesis that regulation operates by switching on and off the synthesis of messenger RNA, it does not by itself constitute a direct proof of the validity of this hypothesis. More direct evidence has, however, been obtained.

First, changes in cultural conditions, which are known either to repress or to induce the synthesis of an important fraction of the total proteins, have been found to result in a detectable decrease, or increase, in the rate of synthesis of messenger RNA (Hiatt *et al.*, 1963). This observation suggests that regulation controls the rate of synthesis of messenger RNA, but it gives no indication as to the specificity of the induced, or repressed, messengers.

For the present time, the only criterion allowing the detection of a specific messenger is the ability of this messenger to form specific hybrids with homologous DNA. In an inducible system, if induction does indeed result in the production of the specific messenger, such a messenger will be found in induced, but not in non-induced bacteria. In other words, formation of hybrid molecules with the corresponding DNA will be observed with messenger from induced, but not from non-induced, bacteria. This type of experiment has been performed with two inducible systems of *E. coli* which are involved in the utilization of lactose and galactose, respectively. In both systems, it is possible to isolate the specific DNA in a reasonably purified form. The results of an experiment involving the system for galactose utilization are reported in Fig. 4. They show unambiguously that the messenger RNA of this system is present only in induced bacteria (Attardi *et al.*, 1962).

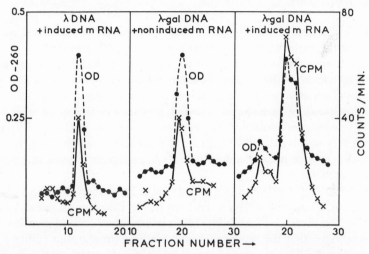

FIG. 4. Induction of specific messenger RNA involved in the production of the enzyme of galactose utilization in *Escherichia coli*. A culture of galactose-positive bacteria (*E. coli* K12, Hfr H) in glycerol minimal medium is divided in two parts: one is grown in the presence of 10^{-3} M fucose (an analog of galactose) which induces the synthesis of the three enzymes of galactose utilization, and the other in the absence of fucose. A pulse of 90 seconds of radiophosphorus is given to the two cultures. Bacteria are harvested in buffer containing 10^{-2} M MgSO$_4$ and ground with alumina. Ribosomes with which messenger RNA is associated are obtained by centrifugation. Total RNA is extracted from ribosomes by phenol extraction and fractionated in sucrose gradient. Suspensions of phages λ and λ-gal (which carries the three bacterial galactose genes) are purified in density gradients of CsCl. DNA from these two phage preparations is extracted by phenol and then denatured by heating followed by fast cooling. Fractions enriched in messenger RNA (from induced or non-induced cells) are heated at 40° C for 4 hours with denaturated DNA (from λ or λ-gal), then slowly cooled to 18° C in order to form DNA-RNA hybrids. After light treatment with RNase (which destroys nonspecific aggregates, but not specific hybrids), the DNA and the hybrids are fractionated in density gradients of CsCl. The ordinates indicate the OD-260 and the counts/min (cpm). The abscissa indicates the fractions of the gradient collected from highest to lowest densities. Messengers from induced cells form a small amount of hybrids with DNA of wild type λ: 53 counts/min/unit OD (*left*). Messengers from noninduced cells form the same amount of hybrids with λ-gal DNA: 51 counts/min/unit OD (*center*). Messengers from induced cells form 3 times more hybrids with λ-gal: 151 counts/min/unit OD (*right*). It is concluded that the specific messengers for the galactose system are produced in detectable amount only when bacteria are grown in the presence of inducer. (From Attardi *et al.*, 1962).

It is therefore concluded that regulation of protein synthesis operates at the level of the genetic material, by switching on or off the production of the primary gene product, messenger RNA.

The Repressor

The last aspect of the model we shall discuss here is concerned with the nature of the repressor and its interaction with specific metabolites. From the previously reported observations, it has been concluded that the repressor—which is defined as the product of the regulator gene, able to exhibit different properties as a result of mutations affecting the regulator gene—acts on the corresponding operator to prevent synthesis of the messengers by the whole operon. In addition, the repressor is able to interact with specific metabolites which can either activate it (repression) or inactivate it (induction). A repressor must therefore be able to recognize a given operator and a given metabolite.

In the case of β-galactosidase (Pardee and Prestidge, 1959) and of phage (Jacob and Campbell, 1959), it has been shown that repression can be established in the presence of inhibitors of protein synthesis, an observation which had suggested that the repressor might not be a protein, but rather a polyribonucleotide. This conclusion, however, if it may explain satisfactorily the specificity of interaction between the repressor and the operator, meets with serious difficulties. (1) The polyribonucleotides, primary products of regulator genes, not being transcribed into protein, would have to be of a different nature from the polyribonucleotides formed by structural genes, i.e., messenger RNA. (2) The recognition of a metabolite by a polynucleotide seems difficult to visualize without the mediation of a protein, which would have to be controlled by the regulator gene. (3) The properties of the different alleles of the regulator genes—and more particularly of the R^s allele which results in a decrease of the affinity of the repressor for the inducer—are difficult to account for if the product of these genes is not a protein.

For the present time, direct identification of the repressor remains a difficult problem. It is possible to demonstrate indirectly that the product of the regulator genes is transcribed into proteins by studying the effects, on regulator mutations, of suppressors known to act at the level of the transcription process from polynucleotidic templates into peptide chains (Yanofsky *et al.*, 1961; Benzer and Champe, 1961). It is clear that if some regulator alleles are found to be susceptible to the action of such supressors, it has to be concluded that the synthesis of the repressor involves a transcription into a peptide chain. This turns out to be the case for the phage repressor. Among 300

R^- mutations of the regulator gene of phage λ, 11 are suppressible by a particular bacterial suppressor, known to restore enzymatic activity of several proteins altered by mutation. Because of its properties of specific steric recognition, the repressor can hardly be a small molecule. It seems, therefore, unlikely that the protein revealed by the effect of the suppressors on the active product of regulator genes is an enzyme controlling the synthesis of the repressor. Most probably, the repressor itself is, partially or in totality, a protein (Jacob *et al.*, 1962).

This conclusion may be reconciled with the fact that repression can be established in the presence of inhibitors of protein synthesis if the number of molecules of a specific repressor per cell is very small. The accumulation of a few regulator messengers in the presence of the inhibitor would, upon removal of the inhibitor, result in an almost immediate synthesis of enough repressor molecules to insure complete repression. In fact, the available evidence, both from kinetic study of induction of β-galactosidase in haploid and diploid bacteria heterozygous for the regulator gene (Ullman *et al.*, 1963) and from the study of immunity in phage (Bertani, 1961; Jacob *et al.*, 1962), suggests that the repressor acts stoichiometrically and points to a very small number of repressor molecules, perhaps not higher than 10 or 15 per chromosome in each case. On the basis of the present knowledge, it appears that the expression of a regulator gene, like that of a structural gene, involves a transcription into protein, the product of a regulator gene being formed in very small amounts.

Furthermore, kinetics of enzyme induction are difficult to reconcile with the hypothesis that the interaction between the repressor and the inducing or repressing, metabolite involves a covalent reaction. The extremely short and constant lag of induction, observed for a wide range of inducer concentration, the almost instantaneous arrest of protein synthesis upon removal of the inducer, as well as the steady rate of enzyme synthesis obtained at a given inducer concentration, whether the previous concentration was higher or lower, all these facts seem to be more compatible with the alternative hypothesis that the *repressor involves an allosteric protein* in the sense previously described. The repressor would possess two distinct sites, one for the operator and one for the specific metabolite. The combination with the metabolite would modify the affinity of the repressor for the operator, either increasing it in the case of enzyme repression, or decreasing it in the case of enzyme induction. Instantaneous switches, on or off, of messenger production would thus result from allosteric transitions of the repressor protein. According to this hypothesis, the whole cellular regulation would ultimately rely on allosterically induced fits or misfits of a few protein species (Ullman *et al.*, 1963).

Genetic Regulation and Differentiation

The study of protein synthesis in bacteria has revealed a system of genetically controlled cytoplasmic signals regulating gene activity. In the bacterial cell, the hereditary message is written in a single linear structure, the bacterial chromosome, which determines the macromolecular pattern of the cell. The transcription of the structural message, written along the chromosome, involves a continuous flow of information from the genetic material to the cytoplasm via metabolically unstable messenger-RNA molecules which bring to the protein-forming centers instructions for building specific protein configurations. There exist, in the chromosome, specialized determinants whose function is to establish specific circuits which, according to the requirements and environmental conditions, allow the selection of the genetic potentialities to be expressed, and therefore of the types of structural messengers produced. At the protein level, other types of circuits interconnecting metabolic pathways regulate enzyme activity by allosteric changes in the configuration of certain key proteins. The different constituents of the bacterial system are thus able to control and inform each other.

It is important to recall, at this point, that the specificity of induction (or repression) as well as of allosteric inhibition in bacterial systems is independent of the specificity of action of the enzymes involved. In the lactose system of *E. coli*, for instance, the synthesis of mutant proteins, which do not exhibit any detectable affinity for β-galactosides, is still normally and specifically induced by these compounds (Perrin *et al.*, 1960). In the same way, we have seen that mutants for the histidine pathway of *Salmonella*, in which the structural genes have been detached from the operator, produce a series of normal enzymes but their synthesis is no longer influenced by histidine. Although inducers are in general substrates, or analogs of the substrate, and repressing metabolites are products (sometimes far removed) of the controlled enzyme, the mechanism of the effect itself imposes no restriction upon the choice of the active agents. The regulation systems may be visualized as "circuits" whose specificity must be considered purely as a result of selection. An allosteric enzyme, or a repressor, is merely a transducer of chemical signals through which interactions occur between reactions which, by virtue of the chemical structure of the reactants, would otherwise proceed independently. Living organisms could not possibly survive, and even less multiply, if it were not for the operation of an exceedingly complex network of regulatory and signaling circuits. The establishment of any sort of complex circuitry, chemical or electric, involves primarily the possibility of interconnecting different parts of the system so that they control and inform each other to the benefit of an adequate final output. Allosteric effects and more

specifically allosteric proteins offer precisely the type of "universal" interconnecting element required for the construction of physiological circuits.

According to our previous definition of differentiation, a bacterial population growing, for instance, in the presence of a specific inducer, and thereby
producing some specific protein(s), is to be considered as differentiated
with respect to a genetically identical population growing in the absence of
inducer. An essential characteristic of regulation of protein synthesis in bacteria, however, is the almost instantaneous response to regulating stimuli and
the complete and instantaneous reversibility observed when the stimuli disappear, as illustrated in Fig. 5. Such a complete reversibility is expected in
unicellular organisms where selection will necessarily favor the most rapid
response to any change of environment. As already pointed out, the specificity of induction or repression of enzyme synthesis in bacteria is not inherently
related to the specificity of the controlled enzyme(s), but merely results
from selection of the most efficient regulatory circuits. If inducing or repressing metabolites in bacteria turns out to be, in general, directly related
to, or identical with, metabolites of the pathway which they control, this
reflects the requirements of free-living unicellular organisms. The problems

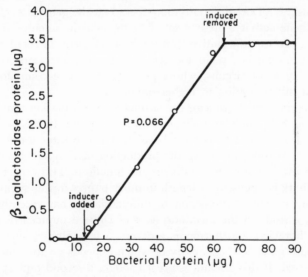

Fig. 5. Kinetics of induced enzyme synthesis. Differential plot expressing accumulation of β-galactosidase as a function of increase of mass of cells in a growing culture of
E. coli. Since abscissa and ordinate are expressed in the same units (micrograms of protein)
the slope of the straight line gives galactosidase as the fraction (P) of total protein synthesized in the presence of inducer. (After Cohn, 1957.)

of such organisms are only to preserve their homeostatic state while adapting rapidly to the chemical challenge of changing environments, and the efficiency of selection depends essentially on a single parameter: the relative rate of multiplication.

The cells of differentiated organisms are faced with entirely different problems. On the one hand, certain groups of functions are permanently delegated to certain groups of cells. On the other hand, intercellular (and not only intracellular) coordination becomes a major factor in selection, while the environment of individual cells is largely stabilized, thus eliminating to a large degree the requirements for rapid and extensive nutritional adaptation.

It is clear, therefore, that differentiated organisms may be expected to possess certain types of regulatory mechanisms which are not found in unicellular organisms. The question, however, is whether cellular regulation and differentiation in higher organisms use the same basic mechanisms as bacterial systems, employing similar circuit elements geared in a different way to meet the requirements of higher organisms. Since bacteria provide us with a model of reversible differentiation operating at the genetic level through circuits of (probably allosteric) signals, we may subdivide the question and ask: (1) Does differentiation operate at the genetic level by turning on or off gene activity, thereby selecting the genetic potentialities to be expressed? (2) In the affirmative, can the basic elements of regulatory circuits found in bacteria, i.e., regulator genes, repressors, operators, be organized into other types of circuits, whose properties could account for the main features of differentiation in higher organisms?

The experimental arguments concerning the first question come from cytogenetical studies of polytenic chromosomes of Diptera. In this material, independent chromosomal units have been shown to exhibit a differential and reversible activation, involving the production of a rapidly renewed RNA, correlated in some instances with specific functions. This aspect of chromosomal activity is discussed at length in other papers in this symposium. The relationship of this phenomenon to differentiation has recently been supported by a study of the metamorphosis of Diptera induced by injection of the prothoracic gland hormone, ecdyson. A few minutes after injection, a specific puff appears, the stability of which depends on the amount of hormone injected. If this amount is great enough, a second puff appears a few minutes later in another chromosomal region, followed by a specific pattern of appearance and disappearance of various puffs in different regions. It is as though the hormone were inducing the formation of a first specific puff which would in turn initiate an orderly series of specific chromosomal events occurring at definite times of metamorphosis (Clever, 1961). All the observations with salivary gland chromosomes strongly suggest that specific modifications of the structure and activity in localized chromosomal regions occur in different tissues of an organism, at different times of the life history of the individual, and that these modifications are reversible and related to the functional activity of the cell.

The second question deals with the mechanisms which may account for the orderly emergence of differentiated functions. The stability and the clonal character of differentiation point to "hereditary" phenomena, but the main problem, which has raised many difficulties of terminology, is concerned with the nature of these phenomena (Nanney, 1958; Ephrussi, 1958; Sonneborn, 1960). As clearly pointed out by Lederberg (1958), this problem must be defined in chemical terms, namely, whether differentiation involves conservation of, or specific changes in, the information coded in the base sequence of polynucleotides. Changes imply any mutational alteration in the sequences, or any distribution of sequences contained in chromosomes or in extranuclear elements. Conservation implies differential functional activity of nucleotide sequences, resulting, for instance, from the establishment of steady state systems capable of clonal perpetuation, as pointed out by Delbrück (1949).

The hypothesis of systematic modifications of the information contained in the genetic material has progressively declined in popularity as knowledge of its structure and functioning increased. Mechanisms involving orderly specific alterations of nucleotide sequences, i.e., mutations, are hardly conceivable at the present time. Although somatic cells have not yet been submitted to genetic analysis, the unequal distribution of chromosomal nucleic information in the course of development appears rather unlikely, except in some rare cases as observed in *Ascaris* or Diptera. The difficulties thus encountered have left hypothetical plasmagenes as the only candidates to account for eventual changes in the content of genetic information during development. Although frequently favored in past years, this hypothesis still lacks the support of experimental evidence as well as of a plausible mechanism to account for an orderly distribution of cytoplasmic particles.

Opposition to the alternative models, i.e., stable activation or inactivation of chromosomal segments, has come mostly from the difficulty of visualizing a suitable mechanism able to alter the function of genes without altering their informational content. The study of genetic regulation in bacteria provides just such a system, or at least the elements of such a system. Like elements of electronic systems, these elements can be organized into a variety of circuits fulfilling a variety of purposes.

Model Circuits

Genetic regulatory circuits, as they occur in bacteria, allow the bacterium to select which of its genetic potentialities have to be expressed according to environmental and cytoplasmic conditions. The bacterial regulatory systems are essentially genetic and negative, i.e., when they are active they prevent the expression of genetic potentialities. Although bacterial circuits are entirely reversible, it is possible to produce other circuits endowed with different degrees of stability, just by connecting the same regulatory elements into other types of circuits (Monod and Jacob, 1961). Such model circuits are, of course, entirely imaginary, but the actual elements of these circuits, namely, regulator genes, repressors, operators, are not imaginary; they are the elements which operate in bacteria.

The system represented in Fig. 6 represents a classical induction system with the peculiarity that the inducer is not the substrate but the product of the controlled enzyme. Such a system, which is known to occur in bacteria, mimics certain properties of genetic elements. In the absence of an exogenous inducer, the enzyme will not be produced, unless already present. When the system is locked, temporary contact with an inducer will unlock it ndefinitely, at least as long as either substrate or product is present.

More complicated circuits can be obtained by interconnecting two inducible, or repressible, systems. In the case represented in Fig. 7, the product of each system acts as a repressing substance for the other. This results in mutually exclusive steady states, the presence of one enzyme blocking permanently the synthesis of the other. A switch from one system to the other is accomplished by temporary elimination of the substrate of the live system. On the contrary, if, as represented in Fig. 8, the product of one enzyme acts as an inducer for the other, the two enzymes are mutually dependent. One could not be synthesized in the absence of the other although they might function in apparently unrelated pathways. Inhibition of one enzyme, or

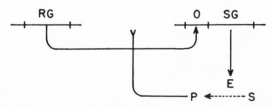

FIG. 6. Model I. Synthesis of enzyme E, genetically determined by the structural gene SG, is blocked by the repressor synthesized by the regulator gene RG. The product P of the reaction catalyzed by enzyme E acts as an inducer of the system by inactivating the repressor. O: operator; S: substrate.

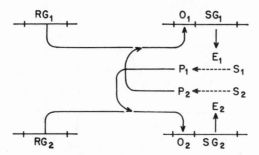

Fig. 7. Model II. Synthesis of enzyme E_1, genetically determined by the structural gene SG_1, is regulated by the regulator gene RG_1. Synthesis of enzyme E_2, genetically determined by the structural gene SG_2, is regulated by the regulator gene RG_2. The product P_1 of the reaction catalyzed by enzyme E_1 acts as corepressor in the regulation system of enzyme E_2. The product P_2 of the reaction catalyzed by enzyme E_2 acts as corepressor in the regulation system of enzyme E_1. O: operator; S: substrate.

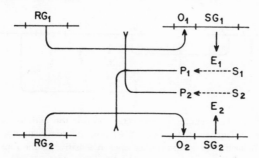

Fig. 8. Model III. Synthesis of enzyme E_1, genetically determined by the structural gene SG_1, is blocked by the repressor synthesized by the regulator gene RG_1. Synthesis of another enzyme E_2, controlled by structural gene SG_2, is blocked by another repressor synthesized by regulator gene RG_2. The product P_1 of the reaction catalyzed by enzyme E_1 acts as an inducer for the synthesis of enzyme E_2, and the product P_2 of the reaction catalyzed by enzyme E_2 acts as an inducer for the synthesis of enzyme E_1. O: operator; S: substrate.

elimination of its substrate, even temporarily, would eventually result in the permanent suppression of both. Finally, the two models can be combined as represented in Fig. 9. Here the product of one enzyme acts as inducer for the second, while the product of the second enzyme exhibits repressing properties for the first enzyme. This is an interesting system because, provided adequate constants are chosen for the decay of the enzymes and their products, the system will oscillate from one state to the other and provide some kind of a biological clock.

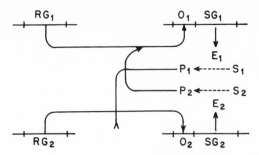

Fɪɢ. 9. Model IV. Synthesis of enzyme E_1, genetically determined by the structural gene SG_1, is blocked by the repressor synthesized by the regulator gene RG_1. Synthesis of another enzyme E_2, controlled by structural gene SG_2, is blocked by another repressor synthesized by regulator gene RG_2. The product P_1 of the reaction catalyzed by enzyme E_1 acts as an inducer for the synthesis of enzyme E_2, while the product P_2 of the reaction catalyzed by enzyme E_2 acts as a corepressor for the synthesis of enzyme E_1. O: operator; S: substrate.

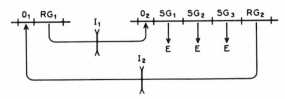

Fɪɢ. 10. Model V. The regulator gene RG_1 controls the activity of an operon containing three structural genes (SG_1, SG_2, SG_3) and another regulator gene RG_2. The regulator gene RG_1 itself belongs to another operon sensitive to the repressor synthesized by RG_2. The action of RG_1 can be antagonized by an inducer I_1, which activates SG_1, SG_2, SG_3 and RG_2 (and therefore inactivates RG_1). The action of RG_2 can be antagonized by an inducer I_2 which activates RG_1 (and therefore inactivates the systems SG_1, SG_2, SG_3 and RG_2). O: operator.

As a last example, we can go one step further and assume that two regulator genes can interact with each other. In the scheme represented in Fig. 10, the regulator gene RG_1 controls the activity of an operon including a second regulator gene RG_2, which in turn can control RG_1. Such a system would be completely independent of the actual metabolic activities of the enzymes. It could be switched from the inactive to the active state by transient contact with a proper inducer, produced, for instance, by another tissue. Once activated, the system could not be switched back except by addition of the repressor produced by the first regulator gene. The change of state would therefore be irreversible. Conversely, starting from the active state, transient contact with an inducer acting on the product of the second regulator gene would switch the system permanently to the inactive state.

These models are not proposed to explain differentiation, but to show how rather simple circuits of negative genetic regulation may mimic very closely changes occurring in the genetic material itself. The transcription of a gene, not only in a cell but in a whole cell lineage, may be permanently repressed— or derepressed—depending on an initial, transient event, which would not depend on a specific alteration of the information carried in the gene.

Other variations of the bacterial model may also result in a greater stability of the circuits. For instance, in bacteria, all the evidence points to a great instability of messenger RNA. In higher organisms, in contrast, some cases are known for which more stable messengers have to be postulated. A rapidly labeled messenger-like RNA has been found in HeLa cells grown in tissue cultures (Scherrer and Darnell, 1962), in the nuclear RNA of thymus cells (Sibatani *et al.*, 1962) and of rat liver cells (Hiatt, 1962), but not in the cytoplasm of liver cells (Hiatt, 1962) or in reticulocytes (Marks *et al.* 1962). Clearly a differential stability of messengers would provide a means for a differential stability of gene expression. Actively growing, nondifferentiated cells might have unstable messengers and constantly reprogrammed ribosomes, while nongrowing, highly differentiated cells might have more stable messengers for the few very particular proteins they make. The messenger-ribosome complex might remain stable so that the ribosomes could not be reprogrammed by other messengers, a system which would lock differentiation in such cells.

Conclusion

In concluding these remarks concerning the possibility of using elements of bacterial circuitry to devise other circuits accounting for the requirements of cellular differentiation, we may turn to the already known example relevant to this discussion.

The existence of regulatory circuits in cells of differentiated organisms is already known. In maize, the work of McClintock (1956) and of Brink (1960) has revealed systems which control gene activity and are essentially composed of two elements. One is closely associated with the structural gene and governs its activity. The other may be located in another region of the chromosomal set. The latter element determines the conditions to which the gene-associated element specifically responds, and therefore the change in the activity of the structural gene. These systems are quite specific: each gene-associated element responds only to a particular second element and it affects only the activity of the gene located in the same chromosome (*cis*). As pointed out by McClintock (1961), such a dual system of control in maize may, in many respects, be compared with the regulator gene-operator system of bacteria. It seems likely that the maize elements also operate by way of specific molecules acting as chemical signals like bacterial repressors.

The generalized hypothesis of genetic regulatory circuits leads to a definite prediction: the possibility of obtaining, in higher organisms, mutants in which the rate of synthesis, but not the structure, of a given protein would be altered. In the synthesis of human hemoglobin, an operator-element has recently been postulated to account for thalassemia and high fetal hemoglobin trait (Neel, 1961; Motulsky, 1962). At birth of a normal child, the production of γ chains, present in fetal hemoglobin, is switched off, whereas production of β and δ chains, present in adult hemoglobin, is switched on. The high fetal hemoglobin trait involves a mutation, extremely closely linked to the structural genes for β and δ chains, which results in the persistence of fetal hemoglobin in otherwise normal adults, heterozygous for this trait. The study of hemoglobin produced by double heterozygotes for this trait and for the structural gene controlling β-chain synthesis shows that the mutation prevents expression of only the immediately adjacent linked β and δ loci of the same (*cis*) chromosome. In the study of haptoglobins, mutations have been found which can hardly be explained by changes in structural genes, but strongly suggest the existence of regulator and operator-like elements controlling the expression of structural genes (Parker and Bearn, 1963). Finally, another example reminiscent of a negative-operator type of mutation is also provided in the cytogenetic studies in salivary glands of *Chironomus*. A specific puff, correlated with the production of certain secretion granules, is observed in one species but not in another, although no difference in the bands of this chromosomal region can be detected. In hybrids between the two species, only that chromosome issued from the puffing parent is found to puff (Beerman, 1961).

All these facts encourage the hypothesis that differentiation operates at the genetic level, using elements basically similar to those found in bacteria, an interpretation which would not be in any way incompatible with the experiments of Briggs and King (1955). As a result of complex interconnections of circuits, successive and orderly switches of elements of the cellular genome, on or off, might provide a precise time clock necessary to control the emergence of functionally differentiated cells within a genetically homogeneous population. In the model of negative genetic regulation, one of the main difficulties which may be expected in a systematic attempt to identify inducers, is that there is no guiding *chemical* principle (based, for instance, on steric analogy). As already discussed for bacterial regulation, the mechanism of the effects imposes no restriction upon the inducing (or repressing) agent interacting with a repression circuit or with an allosteric inhibition. Since the interaction is not limited by, nor dependent upon, any obligatory chemical relationship between substrate and effector, it is clear that virtually *any* physiologically or embryologically useful interconnection between any two or more metabolic pathways, as well as between any two or more tissues or organs, may and indeed should become established by selection of the proper "construction" in the competent system. The freedom of selection prevents us from deducing on a purely chemical basis which type of compound may favor the production of a given enzyme system in the course of development. Hormones appear as obvious candidates for some systems, and several cases are already known of molecular conversions in which a hormone appears to be involved. The case of *Chironomus*, in which ecdyson induces the rapid production of a specific puff, suggests that hormones might also interact with regulatory genetic circuits, although it remains to determine whether the formation of the puff is indeed a primary reaction. In fact, the mode of action of a given chemical substance, such as a hormone, might eventually be multivalent, activating the production of one enzyme system in a certain type of cell, inactivating that of another enzyme system in another type of cell, and acting on the functioning of other enzymes, by allosteric action, in a third type of cell.

The remarkable advances in the methodology of cell cultures are likely to open a new way toward experimental analysis of differentiation. To the present day, the main obstacle has been the impossibility of deciding whether differentiated cells of an organism possess the same structural information coded in their nucleotide sequence. This uncertainty has resulted mainly from the lack of suitable genetic systems allowing genetic analysis of somatic cells. It is not impossible that progress in biochemistry will allow one to bypass in some way genetic analysis by conventional methods. It will probably soon become possible to investigate the specific messengers produced by differentiated cells performing very specialized functions. Whether or not such specialized messengers will recognize the presence of their genes in other cells, by forming hybrid molecules with the DNA of these cells, might perhaps provide a way of analyzing the genetic content of somatic cells. If this content is the same, and if differentiation is based on genetically controlled circuits, then genetic analysis of somatic cells may well turn out to be essentially an analysis of gene expression, as controlled by gene interaction. The main genetic tools would then be provided by the isolation of regulatory mutants, in which specific circuits would be altered, allowing, for instance, the mutant cell to produce molecular species that it does not produce normally. In the analysis of gene interaction, biochemical study may again become a major factor since it now seems reasonable to expect in the near future the production of proteins in subcellular fractions. Reconstructed systems, where DNA or nuclei of certain differentiated cells would be exposed to extracts of other cells, should then provide an assay for the effect of the products of regulatory genes on the expression of structural genes, as well as for a systematic search of inducer substances.

Acknowledgments

The work done in the authors' laboratory has been supported by grants from the National Science Foundation, The Jane Coffin Memorial Fund, and the Commissariat à l'Energie Atomique.

References

Algranati, I. D., and Cabib, E. (1962). Uridine diphosphate D-glucose-glycogen glucosyltransferase from yeast. *J. Biol. Chem.* **237**, 1007–1013.

Ames, B. N., and Garry, B. (1959). Coordinate repression of the synthesis of four histidine biosynthetic enzymes by histidine. *Proc. Natl. Acad. Sci. U.S.* **45**, 1453–1461.

Ames, B. N., Garry, B. and Herzenberg, L. A. (1960). The general control of enzymes of histidine biosynthesis in *Salmonella typhimurium*. *J. Gen. Microbiol.* **22**, 369–378.

Ames, B. N., Hartman, P. E., and Jacob, F. (1963). Chromosomal alterations affecting the regulation of histidine biosynthetic enzymes in *Salmonella*. *J. Mol. Biol.* in press.

Attardi, G., Naono, S., Gros, F., Brenner, S., and Jacob, F. (1962). Effet de l'induction enzymatique sur le taux de synthèse d'un ARN messager spécifique chez *E. coli*. *Compt. rend. acad. sci.* **255**, 2303–2305.

Beckwith, J. R., Pardee, A. B., Austrian, R., and Jacob, F. (1962). Coordination of the synthesis of the enzymes in the pyrimidine pathway of *E. coli*. *J. Mol. Biol.* **5**, 618–634.

Beerman, W. (1961). Ein Balbiani-Ring als Locus einer Speicheldrüsenmutation. *Chromosoma* **12**, 1–25.

Benzer, S., and Champe, S. P. (1961). Ambivalent *rII* mutants of phage T_4. *Proc. Natl. Acad. Sci. U.S.* **47**, 1025–1038.

Bertani, L. E. (1961). Levels of immunity to superinfection in lysogenic bacteria as affected by prophage genotype. *Virology* **13**, 378–380.

Braun, A. C. (1961). Plant tumors as an experimental model. *Harvey Lectures* **56**, 191–210.

Brenner, S., Jacob, F., and Meselson, M. (1961). An unstable intermediate carrying information from genes to ribosomes for protein synthesis. *Nature* **190**, 576–581.

Briggs, R. W., and King, T. J. (1955). *In* "Biological Specificity and Growth," Growth Symposium No. 12 (E. G. Butler ed.), pp. 207–228. Princeton Univ. Press, Princeton, New Jersey.

BRINK, R. A. (1960). Paramutation and chromosome organization. *Quart. Rev. Biol.* **35**, 120–137.

BUTTIN, G. (1962). Sur la structure de l'opéron galactose chez *E. coli. Compt. rend. acad. sci.* **255**, 1233–1235.

CHAMBERLIN, M., AND BERG, P. (1962). Deoxyribonucleic acid-directed synthesis of ribonucleic acid by an enzyme from *Escherichia coli. Proc. Natl. Acad. Sci. U.S.* **48**, 81–94.

CHANGEUX, J. P. (1961). The feedback control mechanism of biosynthetic L-threonine deaminase by L-isoleucine. *Cold Spring Harbor Symposia Quant. Biol.* **26**, 313–318.

CHANGEUX, J. P. (1962). Effet des analogues de la L-thréonine et de la L-isoleucine sur la L-thréonine désaminase. *J. Mol. Biol.* **4**, 220–225.

CLEVER, U. (1961). Genaktivitäten in den Riesenchromosomen von Chironomus Tentans und ihre Beziehungen zur Entwicklung. *Chromosoma* **12**, 607–675.

COHN, M. (1957). Contributions of studies on the β-galactosidase of *Escherichia coli* to our understanding of enzyme synthesis. *Bacteriol. Revs.* **21**, 140–168.

DELBRÜCK, M. (1949). *In* "Unités biologiques douées de continuité génétique," pp. 33–34. C.N.R.S., Paris.

DEMEREC, M., AND HARTMAN, P. E. (1959). Complex loci in microorganisms. *Ann. Rev. Microbiol.* **13**, 377–406.

EPHRUSSI. B. (1958). The cytoplasm and somatic cell variation. *J. Cellular Comp. Physiol.* **52**, 35–53.

FRISCH, L., ed. (1961). Cellular regulatory mechanisms. *Cold Spring Harbor Symposia Quant. Biol.* **26**.

GERHART, J. C. AND PARDEE, A. B. (1962). The enzymology of control by feedback inhibition. *J. Biol. Chem.* **237**, 891–896.

GORINI, L., GUNDERSEN, W., AND BURGER, M. (1961). Genetics of regulation of enzyme synthesis in the arginine biosynthetic pathway of *Escherichia coli. Cold Spring Harbor Symposia Quant. Biol.* **26**, 173–182.

GROS, F., GILBERT, W., HIATT, H., KURLAND, C. G., RISEBROUGH, R. W., AND WATSON, J. D. (1961a). Unstable ribonucleic acid revealed by pulse labelling of *Escherichia coli. Nature* **190**, 581–585.

GROS, F., GILBERT, W., HIATT, H., ATTARDI, G., SPAHR, P. F., AND WATSON, J. D. (1961b). Molecular and biological characterization of messenger RNA. *Cold Spring Harbor Symposia Quant. Biol.* **26**, 111–132.

HALL, B. D., AND SPIEGELMAN, S. (1961). Sequence complementarity of T2-DNA and T2-specific RNA. *Proc. Natl. Acad. Sci. U.S.* **47**, 137–146.

HARTMAN, P. E., LOPER, J. C., AND SERMAN, D. (1960). Fine structure mapping by complete transduction between histidine-requiring *Salmonella* mutants. *J. Gen. Microbiol* **22**, 323–353.

HIATT, H. H. (1963). A rapidly labeled RNA in rat liver nuclei. *J. Mol. Biol.* **5**, 217–229.

HIATT, H. H., GROS, F., AND JACOB, F. (1962). The effect of induction and repression on the rate of synthesis of messenger ribonucleic acid. *Biochim. et Biophys. Acta*, in press.

HORIUCHI, T., HORIUCHI, S., AND NOVICK, A. (1961). A temperature-sensitive regulatory system. *J. Mol. Biol.* **3**, 703–704.

HURWITZ, J., FURTH, J. J., ANDERS, M., ORTIZ, P. J., AND AUGUST, J. T. (1961). The enzymatic incorporation of ribonucleotides into RNA and the role of DNA. *Cold Spring Harbor Symposia Quant. Biol.* **26**, 91–100.

JACOB, F., AND CAMPBELL, A. (1959). Sur le système de répression assurant l'immunité chez les bactéries lysogènes. *Compt. rend. acad. sci.* **248**, 3219–3221.

JACOB, F., AND MONOD, J. (1961a). Genetic regulatory mechanisms in the synthesis of proteins. *J. Mol. Biol.* **3**, 318–356.

JACOB, F., AND MONOD, J. (1961b). On the regulation of gene activity. *Cold Spring Harbor Symposia Quant. Biol.* **26**, 193–211.

JACOB, F., SUSSMAN, R., AND MONOD, J. (1962). Sur la nature du répresseur assurant l'immunité des bactéries lysogènes. *Compt. rend. acad. sci.* **254**, 4214–4216.

KOSHLAND, D. E., JR. (1959). Enzyme flexibility and enzyme action. *J. Cellular Comp. Physiol.* **54**, 245–258.

LEDERBERG, J. (1958). Genetic approaches to somatic cell variation: Summary comment. *J. Cellular Comp. Physiol.* **52**, 383–401.

LEE, N., AND ENGLESBERG, E. (1962). Dual effects of structural genes in *Escherichia coli. Proc. Natl. Acad. Sci. U.S.* **48**, 335–348.

LELOIR, L. F., AND CARDINI, C. E. (1957). Biosynthesis of glycogen from uridine diphosphate glucose. *J. Am. Chem. Soc.* **79**, 6340–6341.

LENGYEL, P., SPEYER, J. F., AND OCHOA, S. (1961). Synthetic polynucleotides and the amino acid code. *Proc. Natl. Acad. Sci. U.S.* **47**, 1936–1942.

MAAS, W. K. (1961). Studies on repression of arginine biosynthesis in *Escherichia coli. Cold Spring Harbor Symposia Quant. Biol.* **26**, 183–191.

McCLINTOCK, B. (1956). Controlling elements and the gene. *Cold Spring Harbor Symposia Quant. Biol.* **21**, 197–216.

McCLINTOCK, B. (1961). Some parallels between control systems in maize and in bacteria. *Am. Naturalist* **95**, 265–277.

MARKS, P. A., WILLSON, C., KRUH, J., AND GROS, F. (1962). Unstable ribonucleic acid in mammalian blood cells. *Biochem. Biophys. Research Communs.* **8**, 9–14.

MARTIN, R. G. (1963). The first enzyme in histidine biosynthesis: The nature of feedback inhibition by histidine. *J. Biol. Chem.* in press.

MATTHAEI, J. F., JONES, O. W., MARTIN, R. G., AND NIRENBERG, M. W. (1962). Characteristics and composition of RNA coding units. *Proc. Natl. Acad. Sci. U.S.* **48**, 666–677.

MONOD, J., AND JACOB, F. (1961). General conclusions: Teleonomic mechanisms in cellular metabolism, growth, and differentiation. *Cold Spring Harbor Symposia Quant. Biol.* **26**, 389–401.

MONOD, J., JACOB, F., AND GROS, F. (1961). Structural and rate-determining factors in the biosynthesis of adaptive enzymes. *Biochem. Soc. Symposia (Cambridge, Engl.).* **21**, 104–132.

MOTULSKY, A. G. (1962). Controller genes in synthesis of human haemoglobin. *Nature* **194**, 607-609.

NANNEY, D. L. (1958). Epigenetic control systems. *Proc. Natl. Acad. Sci. U.S.* **44**, 712–717.

NEEL, J. V. (1961). The hemoglobin genes: A remarkable example of the clustering of related genetic functions on a single mammalian chromosome. *Blood* **18**, 769.

NIRENBERG, M. W., AND MATTHAEI, J. H. (1961). The dependence of cell-free protein synthesis in *E. coli* upon naturally occurring or synthetic polyribonucleotides. *Proc. Natl. Acad. Sci. U.S.* **47**, 1588–1602.

NOVICK, A., AND SZILARD, L. (1954). Experiments with the chemostat on the rates of amino acid synthesis in bacteria. *In* "Dynamics of Growth Processes," Growth Symposium No. 11 (E. J. Boell, ed.), p. 21. Princeton Univ. Press, Princeton, New Jersey.

PARDEE, A. B., AND PRESTIDGE, L. S. (1959). On the nature of the repressor of β-galactosidase synthesis in *Escherichia coli*. *Biochim et Biophys. Acta* **36**, 545–547.

PARKER, W. C., AND BEARN, A. G. (1963). In press.

PERRIN, D., JACOB, F., AND MONOD, J. (1960). Biosynthèse induite d'une protéine génétiquement modifiée ne présentant pas d'affinité pour l'inducteur. *Compt. rend. acad. sci.* **251**, 155–157.

RALL, T. W., AND SUTHERLAND, E. W. (1961). The regulatory role of adenosine-3′,5′-phosphate. *Cold Spring Harbor Symposia Quant. Biol.* **26**, 347–354.

RILEY, M., PARDEE, A. B., JACOB, F., AND MONOD, J. (1960). On the expression of a structural gene. *J. Mol. Biol.* **2**, 216–225.

SCHERRER, K., AND DARNELL, J. E. (1962). Sedimentation characteristics of rapidly labeled RNA from Hela cells. *Biochem. Biophys. Research Communs.* **7**, 486–490.

SIBATANI, A., DE KLOET, S. R., ALLFREY, V. G., AND MIRSKY, A. E. (1962). Isolation of a nuclear RNA fraction resembling DNA in its base composition. *Proc. Natl. Acad. Sci. U.S.* **48**, 471–477.

SONNEBORN, T. M. (1960). The gene and cell differentiation. *Proc. Natl. Acad. Sci. U.S.* **46**, 149–165.

STADTMAN, E. R., COHEN, G. N., LE BRAS, G., AND DE ROBICHON-SZULMAJSTER, H. (1961). Selective feedback inhibition and repression of two aspartokinases in the metabolism of *Escherichia coli*. *Cold Spring Harbor Symposia Quant. Biol.* **26**, 319–321.

STEVENS, A. (1960). Information of the adenine ribonucleotide into RNA by cell fractions from *E. Coli B. Biochem Biophys. Research Communs.* **3**, 92–96.

SUSSMAN, R., AND JACOB, F. (1962). Sur un système de répression thermosensible chez le bactériophage λ d'*Escherichia coli*. *Compt. rend. acad. sci.* **254**, 1517–1519.

TISSIERES, A., AND HOPKINS, J. W. (1961). Factors affecting amino acid incorporation into proteins by *Escherichia coli* ribosomes. *Proc. Natl. Acad. Sci. U.S.* **47**, 2015–2023.

TISSIERES, A., SCHLESSINGER, D., AND GROS, F. (1960). Amino acid incorporation into proteins by *Escherichia coli* ribosomes. *Proc. Natl. Acad. Sci. U.S.* **46**, 1450–1463.

TOMKINS, G. M., AND YIELDING, K. L. (1961). Regulation of the enzymic activity of glutamic dehydrogenase mediated by changes in its structure. *Cold Spring Harbor Symposia Quant. Biol.* **26**, 331–341.

TRAUT, R. R. (1962). Glycogen synthesis from UDPG. Thesis, Rockefeller Institute, New York.

ULLMAN, A., MONOD, J., AND JACOB, F. (1963). In preparation.

UMBARGER, H. E. (1956). Evidence for a negative feedback mechanism in the biosynthesis of isoleucine. *Science* **123**, 848.

UMBARGER, H. E. (1961). Feedback control by endproduct inhibition. *Cold Spring Harbor Symposia Quant. Biol.* **26**, 301–312.

VOLKIN, E., AND ASTRACHAN, L. (1957). In "The Chemical Basis of Heredity" (W. D. McElroy and B. Glass, eds.) pp. 686–694. Johns Hopkins Press, Baltimore, Maryland.

WEISS, S. B., AND NAKAMOTO, T. (1961). Net synthesis of ribonucleic acid with a microbial enzyme requiring deoxyribonucleic acid and four ribonucleoside triphosphates. *J. Biol. Chem.* **236**, PC18–20.

WILLSON, C., COHN, M., PERRIN, D., JACOB, F., AND MONOD, J. (1963). in preparation.

WOOD, W. B., AND BERG, P. (1962). The effect of enzymatically synthesized ribonucleic acid on amino acid incorporation by a soluble protein-ribosome system from *Escherichia coli*. *Proc. Natl. Acad. Sci. U.S.* **48**, 94–104.

YANOFSKY, C., HELINSKI, D. R., AND MALING, B. D. (1961). The effects of mutation on the composition and properties of the A protein of *Escherichia coli* tryptophan synthetase. *Cold Spring Harbor Symposia Quant. Biol.* 26, 11–24.

YATES, R. A., AND PARDEE, A. (1956). Control of pyrimidine biosynthesis in *Escherichia coli* by a feed-back mechanism. *J. Biol. Chem.* 221, 757–769.

YČAS, M., AND VINCENT, W. S. (1960). A ribonucleic acid fraction from yeast related in composition to deoxyribonucleic acid. *Proc. Natl. Acad. Sci. U.S.* 46, 804–810.

Epigenetic Control of Specific Protein Synthesis in Differentiating Cells

CLEMENT L. MARKERT

Department of Biology, The Johns Hopkins University, Baltimore, Maryland

The recognition of the chromosomal basis of inheritance some 60 years ago immediately posed a dilemma which has not yet been resolved and which in fact sums up what may be the fundamental problem of embryonic development. It was observed that during cell division each daughter cell received an identical set of chromosomes, presumably therefore an identical set of hereditary potentialities. Yet these daughter cells commonly followed their own independent pathways of development until finally at late stages of differentiation they exhibited very different phenotypes. Thus the dilemma: identical genotypes give rise to very different phenotypes—as different as nerve, and muscle, and pigment cells. With the tremendous expansion of our knowledge of genetics and biochemistry, of the structure and function of genes, of deoxyribonucleic acid (DNA), ribonucleic acid (RNA), and protein, and of their complementary structural and chemical relationships, this basic dilemma has become sharper and its solution so much the more important for our understanding of cell differentiation and embryonic development.

We know now that the metabolic machinery of a cell is largely regulated by enzymes and that the structure of enzymes, like other proteins, is ultimately encoded in the DNA of the chromosomes and probably also in RNA. So far as we can presently judge, the cells of a metazoan have constant identical supplies of DNA, but their enzymatic composition varies enormously. In fact the enzymatic content of a cell is the principal feature by which we assess the biochemical state of differentiation of the cell. Clearly some mechanism must regulate gene expression and this mechanism, whatever it may be, should provide the key to our dilemma and simultaneously elucidate the most basic problems of development. In searching for a mechanism that could selectively and differentially control gene expression in terms of the enzymes produced, we might profitably examine each step leading to the synthesis of an enzyme from the DNA to messenger RNA to ribosome to released enzyme. A limitation imposed anywhere along this sequence could regulate the enzymatic repertory of a cell and thus specify its state of differentiation. However, the problems of control seem to enlarge rapidly as the distance from the chromosome increases.

At the level of the chromosome we can easily imagine a single molecule activating or inhibiting a gene, turning it on or off, as it were. But at all later steps many molecules in many locations would be required to stifle the enzymatic expression of a functioning gene. Effective control of the enzyme molecules themselves would seem to involve enormous logistic problems and constitute a profligate drain on the resources of the cell, but then, perhaps the cell does not view economy the way we do. It should be remembered that the titer of any given enzyme per cell may range from many millions of molecules to none so far as we can tell, or at least to a number so low as to have no physiological significance for the cell. Turning the gene off seems to be the most satisfactory method of completely preventing the appearance of a particular enzyme in the cell. But what of the problem of regulating the amount of enzyme produced once the gene is turned on? Again the problem seems easiest to solve at the gene itself, but we have few facts to guide us in constructing a hypothesis of molecular regulation of the quantity of gene function. Taking a cue from the exciting work on the regulation of bacterial genes summarized by Jacob and Monod (1963), we might assume that the quantitative control of gene function is determined by the duration of gene activation which in turn is a function of the abundance of unstable activating molecules. The more abundant the activator, the greater would be the time during which the activator and gene were associated, and thus the more numerous would be the gene products.

In our efforts to formulate a mechanism for regulating gene function, it is important to bear in mind that the mechanism cannot be an autonomous expression of the genome of any particular cell. The mechanism is clearly dependent upon external influences. Transplanting an embryonic cell from one tissue environment to another will commonly change the fate of the transplanted cell to conform with its new location. Thus gene expression is dependent upon the extracellular environment, although more immediately upon the environment within the cell. Obviously a cyclical, dynamic interaction between the genome and its environment occurs, so that the environment specifies which part of the genome is to function and to what degree, and the functioning genome in turn modifies the environment. This reciprocal interplay of genome and environment drives the cell along the path of differentiation and is basically responsible for embryonic development.

Perhaps the most important, and certainly the most neglected, portion of this cycle is the chromosomes themselves—particularly the structural and chemical changes which occur in them as their function changes (Gall and Callan, 1962; Beermann, 1961; Clever, 1961). Our knowledge of chromosome change as related to function is still meager, but other parts of this cycle of interaction, particularly the synthesis of specific proteins, can more profitably be reviewed today. The mechanisms—genetic and epigenetic—that bring about the synthesis of protein molecules can best be studied by focusing on a specific protein. The enzyme lactate dehydrogenase (LDH) has been extensively investigated by many laboratories, and now provides a rich source of information for analyzing problems of protein synthesis. This oxidoreductase is ubiquitous in vertebrate tissues and is found in many other organisms as well. It catalyzes the interconversion of pyruvate and lactate and simultaneously of $NADH_2$ and NAD.* Lactate appears to have little metabolic importance other than as a temporary storage reservoir for hydrogen during periods of relative anaerobiosis. Pyruvate, of course, occupies a key position in carbohydrate metabolism.

Isozymes

From our understanding of the gene control of enzyme synthesis, the encoding of a single protein by a single gene would lead us to suppose that all the molecules of LDH within a single individual should be identical, except possibly for rare accidents. We were surprised, therefore, a few years ago (Markert and Møller, 1959), to find that LDH exists in several distinct molecular varieties, five in fact, in the tissues of vertebrates. These multiple molecular forms, or isozymes, of LDH, present in a sharp fashion the problems involved in the genetic and also in the epigenetic control of specific protein synthesis in differentiating cells.

The fact that LDH exists in more than one molecular variety in a single organism was first recognized by Meister (1950) and later confirmed by Neilands (1952). Using zone electrophoresis, Wieland and Pfleiderer (1957) and Pfleiderer and Jeckel (1957) were later able to demonstrate additional molecular varieties of LDH and to show that they existed in tissue- and species-specific patterns. Other enzymes have also been shown to exist in multiple molecular forms, but generally the early observations were considered, at best, to be biochemical curiosities or, at worst, evidence of defective preparative procedures leading to artifacts in the form of partially degraded but still active enzyme molecules. A full appreciation of the fact

---

* NAD = nicotinamide-adenine dinucleotide, replacing DPN (diphosphopyridine nucleotide); $NADH_2$ = dihydronicotinamide-adenine dinucleotide, replacing DPNH (diphosphopyridine nucleotide, reduced form).

that many enzymes normally exist in several different isozymic forms depended upon the development of analytical procedures that permitted a direct assay of the isozymic content of tissues with a minimum of manipulation and without the use of conventional preparative procedures.

The development of techniques (Smithies, 1955) for the resolution of protein mixtures by electrophoresis in starch gels provided the basic analytical procedure. The employment of histochemical staining techniques to visualize enzymes separated on starch gels (Hunter and Markert, 1957) provided a simple procedure for direct analysis of enzymes in tissue homogenates. By these methods the isozymes of numerous enzymes, particularly those of LDH, were readily separated and identified (Markert and Møller, 1959). These stained starch strips (Fig. 1), or zymograms, revealed the existence of five isozymes of LDH in all the mammals so far examined (Fig. 2).

The common occurrence of the same number of LDH isozymes in different mammals suggested that the isozymes probably had biological significance. This conclusion became obvious when a variety of different tissues were analyzed. Nearly every tissue contained all five isozymes of LDH in precise relative proportions—that is, the zymogram of each tissue showed a characteristic constant pattern of isozymes. These patterns are very stable and are not easily modified by procedures commonly employed in the purification of enzymes. Homogenizing different tissues together resulted in zymograms that showed merely the summation of the patterns of the constituent tissues.

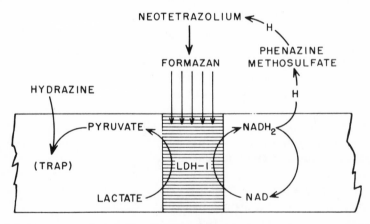

Fig. 1. Diagram of the reactions involved in visualizing the isozymes of LDH after resolution by starch gel electrophoresis. The method essentially provides a colorimetric test for the production of $NADH_2$.

(+)

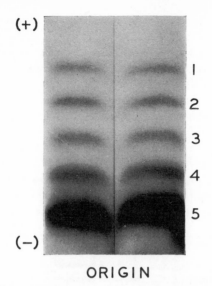

1

2

3

4

5

(−)

ORIGIN

Fig. 2. Photograph of zymograms of two independent preparations of mouse LDH from skeletal muscle. Note that all five isozymes are present. Electrophoresis occurred at pH 7.0 at room temperature for about 6 hours at a voltage gradient of 6 V/cm.

Neither inhibition nor activation of individual isozymes occurréd. Moreover, during the several steps involved in purification the isozymic pattern remains essentially constant. From these observations it seems clear that isozyme patterns, as seen on zymograms, reflect conditions within the tissue and are definitely not artifacts of the analytic procedures.

Since these analyses were performed on tissues that were, of course, composed of heterogeneous populations of cells, it seemed possible that the pattern of isozymes observed represented a corresponding pattern of cell types, each contributing a single isozyme to the total. Two lines of investigation provided at least a partial answer to this possibility. First, a single cell type—mouse erythrocytes—was obtained in pure form. These erythrocytes were separated from the cells of the blood by centrifugation, washed thoroughly in physiological saline solution, lysed in distilled water, and the resulting solution electrophoresed. This single cell type produced a complex pattern of several isozymes. Thus, one might conclude that a single cell can synthesize more than one isozyme, although these observations do not preclude the possibility that the erythrocytes themselves are a heterogeneous population in terms of isozymic content. This reservation seems unlikely in view of the work of Nace et al. (1961). By using fluorescent labeled antibodies to different LDH isozymes of the frog, these investigators were able to demonstrate the presence of three isozymes in the egg of the frog *Rana pipiens*. It seems probable, therefore, that single cells do produce more than one isozyme. However, the diverse cells composing a tissue do not make identical contributions to the isozyme pattern of the tissue. This is easily demonstrated by dividing a complex organ, such as the stomach, into distinct parts and analyzing each of these separately. Quite different isozyme patterns are commonly obtained from these different parts of an organ. The pattern of any organ or tissue is merely the summation of the patterns of its constituent parts.

It is obvious from an examination of the zymograms (Fig. 3) of the different tissues of the mouse that the specific characteristics of each tissue or organ is based upon the specific proportions of the several isozymes present

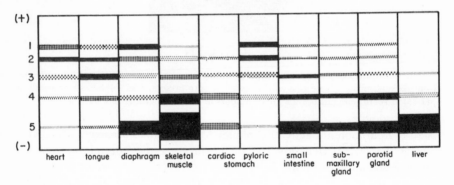

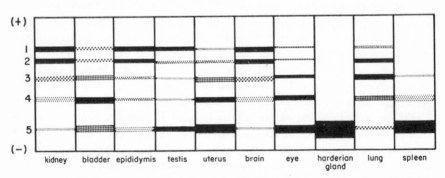

Fig. 3. Diagrammatic representation of the LDH zymogram patterns from 20 tissues or organs of the adult mouse. The darkness of the isozyme bands is a rough indication of the amount of enzyme activity in the band. (From *Develop. Biol.*, **5**, 363–381.)

in the tissue and not upon the presence or absence of an isozyme (Markert and Ursprung, 1962). The concept of a tissue-specific isozyme is entirely inappropriate. Furthermore, all of the isozymes appear to be equally important. Although LDH-5 predominates in many mouse tissues, each isozyme is the most abundant in some tissue. Nearly all tissues contain measurable amounts of all five isozymes, and the few exceptions probably represent cases in which the titer of the isozyme is too low to be detected by these techniques. The range of variation on a zymogram from the faintest to the darkest band represents about a hundredfold difference in isozyme titer.

The fact that each cell can synthesize more than one isozyme raises the question as to where these isozymes may be located within the cell. Reliance must be placed on cytochemical techniques for identifying the location of enzymes within cells. Such techniques are difficult to apply to LDH because of the ready solubility of this enzyme. However, some progress has been made. Allen (1961) demonstrated that LDH is not uniformly distributed throughout the cell and, in fact, different cell types show characteristic distributions of the LDH. His results did not permit the identification of the different isozymic forms of LDH but only located the enzymatic activity. However, the use of fluorescent labeled antibodies (Nace *et al.*, 1961) against individual isozymes did permit localization to individual cells, and even intracellular location could be ascertained with some accuracy, particularly in large cells such as the frog egg. Nace and his associates eluted fluorescent labeled antibodies from precipitin bands in Ouchterlony plates. With these antibodies they then stained tissue sections and demonstrated the differential location of individual isozymes. For example, LDH-1 was identified on yolk platelets of the egg and in the connective tissue of the oviduct. LDH-3 was found in the egg cytoplasm and in certain cells of the oviduct epithelium and in jelly-secreting glands of the oviduct. The location of LDH-2 in the egg was not determined but it was found in certain cells of the oviduct epithelium. Walker and Seligman (1963) have also described precise intracellular localizations of LDH activity which were characteristic for different cell types, although the techniques used by these investigators do not resolve the individual isozymes. All these data suggest that each isozyme may be located in a prescribed site in the cell. Presumably the difference in net charge by which we are able to separate the isozymes electrophoretically is an important molecular property in allowing each isozyme to be held at specific sites in the cell.

The specificity of isozyme distribution in the various tissues of vertebrates strongly implies biological significance. Moreover, since the patterns of adult tissues are different, it follows that these patterns must have arisen during the

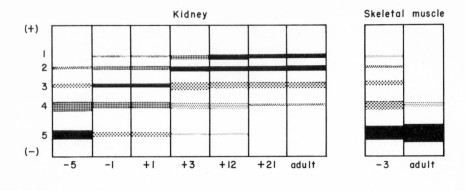

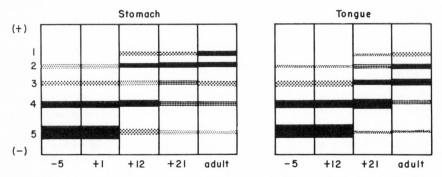

FIG. 4. Diagrammatic representation of changing LDH patterns during the development of several mouse tissues. Note that patterns do not change synchronously. The numbers along the abscissa indicate days before (−) or after (+) birth. (From *Develop. Biol.*, **5**, 363–381.)

course of embryonic development. Direct analysis of tissues at successive steps in development shows this to be so. The adult patterns gradually emerge and in so doing exhibit remarkable regularities of change from the embryonic precursor patterns (Fig. 4). In the mouse, the most extensively investigated mammal, the largest part of the pattern change occurs during early neonatal existence, although progressive changes also occur during embryonic life. Once adulthood is reached, no further change is observed, even in very old mice. All embryonic tissues first exhibit a predominance of LDH-5. As development proceeds a gradual shift in pattern occurs so that enzyme activity is progressively transferred toward the LDH-1 end of the spectrum. The extent of this shift varies enormously in different tissues. In skeletal muscle, for example, very little change in pattern occurs, nearly all

LDH activity being in LDH-5, both in the embryo and in the adult. Heart muscle exhibits quite another picture. In this tissue enzyme activity is progressively shifted from LDH-5 toward LDH-1. In the adult heart nearly all LDH activity is found in isozymes LDH-1 and LDH-2 with almost none remaining in the previously dominant LDH-5. Yet other tissues show intermediate degrees of change from the embryonic emphasis on LDH-5 to the increased abundance of LDH-1 generally characteristic of adult tissues.

This progressive transposition of enzyme activity from one end of the spectrum of isozymes to the other does not occur synchronously throughout the developing mouse. Some tissues mature faster than others so far as their isozyme patterns are concerned. At the same stage of development, for example, the kidney shows a more nearly adult pattern than does the heart, even though the heart functions in the embryo much as it does in the adult. Each tissue evidently matures at its own characteristic rate, largely independently of other tissues.

The pattern changes characteristic of mouse development are paralleled in the development of the chick (Lindsay, 1962) but with certain conspicuous and instructive exceptions. LDH-5 is not the principal isozyme in the chick embryo as it is in the mouse, but rather LDH-1 first appears. In certain tissues such as breast muscle, LDH activity is gradually shifted along the spectrum of isozymes until in the adult only LDH-5 is detected. No change occurs in the chick heart pattern; LDH-1 remains predominant throughout life. Thus adult patterns in the chick are similar to those of the adult mouse, but the starting patterns in the embryo are quite different and therefore the sequence of pattern change is also different. As will be discussed later these differences between chick and mouse are probably related to the availability of oxygen during embryonic development.

Physicochemical Nature of LDH

A full appreciation of the biological significance of isozymes depends upon a detailed knowledge of the physical and chemical differences among them. Accordingly we have undertaken to analyze the physicochemical nature of the LDH isozymes (Markert and Appella, 1961). Highly purified, crystalline preparations of LDH have been made from beef, pig, mouse, and chicken muscle. The most extensive analysis has been made with beef LDH (Table I) but the same general results also seem to apply to LDH from the other organisms.

Beef LDH has a molecular weight of 135,000 as determined from measurements with the ultracentrifuge and by light scattering. In the ultracentrifuge crystalline preparations from beef skeletal muscle, containing all five iso-

TABLE I
PHYSICOCHEMICAL PROPERTIES OF LDH ISOZYMES[a]

Physicochemical property	LDH-1	LDH-5
Molecular weight	135,000	135,000
Sedimentation coefficient ($S_{20,w}^{0} \times 10^{-13}$ cm/sec)	7.0	7.0
Diffusion coefficient ($D_{20,w} \times 10^{-7}$ cm²/sec)	5.10	5.10
Partial specific volume (W_{20} ml/gm)	0.750	0.750
Isoelectric point	4.5	9.5
Subunit size	35,000	35,000
Electrophoretic mobility (pH 7.2, 0.1 ionic strength phosphate buffer)	−4.90	+0.63

[a] From beef tissues.

zymes, show a single symmetric peak. This monodisperse behavior in the ultracentrifuge suggests that all five isozymes are the same size. By contrast, free boundary electrophoresis or zone electrophoresis in starch gels reveals five distinct molecular species. Thus the net charge on each isozyme is different. At pH 7.0 the isozymes are equally spaced along the starch gel after electrophoresis. This arrangement suggests that each isozyme differs from the next in the series by the same increment of charge. A plausible molecular basis for such regularity will be discussed later. Each of these isozymes may be separated from the crystalline preparation by electrophoresis in a column of cellulose powder followed by elution with buffer. Such purified isozymes have the same molecular weight as the crystalline preparations containing several isozymes. Thus the possibility that isozymes are polymers is excluded. They are all the same size. They are not equally stable, however. The curves depicting the course of denaturation by heat for beef LDH-1 and LDH-5 reveal that LDH-5 is noticeably less stable. This fact suggests that the tertiary structure of the two molecules might be significantly different. The tertiary structure relies heavily upon hydrogen bonds to hold the molecule in a stable configuration. Agents that rupture hydrogen bonds might then reveal differences among isozymes if indeed the differences lie in the tertiary structure.

Subunit Hypothesis

When hydrogen bonding reagents such as 12 M urea and 5 M guanidine hydrochloride were added to solutions of the isozymes, they were readily denatured; all enzymatic activity disappeared, as was expected. However, more important was the fact that the molecule was split into four polypeptide chains of equal size as shown by measurements in the ultracentrifuge (Fig.

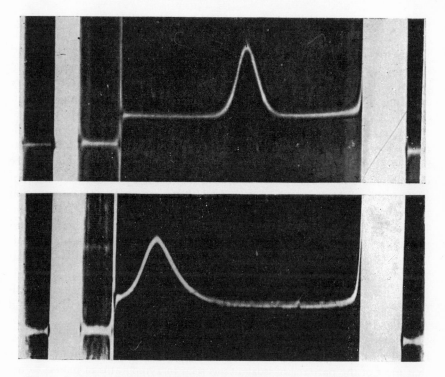

Fɪɢ. 5. Schlieren photographs taken at the same time after the beginning of ultra-centrifugation. Upper photo is of intact LDH preparation; lower photo is after dissociation of LDH into subunits. The difference in sedimentation constants indicates that the subunits are one-fourth the size of the intact molecule. Note that each preparation is monodisperse.

5). Low pH will also dissociate the LDH molecule into its constituent poly-peptides. These polypeptides have no enzyme activity and may be separated into at least two electrophoretically distinct forms. The subunits obtained from LDH-1 and from LDH-5 each appear to be electrophoretically homogeneous but different from each other. The subunits of LDH-2, -3, and -4 all appear to be mixtures of the two kinds of subunits found in LDH-1 and -5. Complex crystalline preparations containing several isozymes also dissociate into two kinds of subunits. Assorting these two kinds of subunits in all possible combinations of four would yield five distinct molecular varieties (Appella and Markert, 1961). It is surely more than coincidence that five LDH isozymes are found in nearly all mammalian tissues so far examined. If the two subunits are designated A and B, then the formulas for the five isozymes can be written A^0B^4, (LDH-1); A^1B^3 (LDH-2); A^2B^2 (LDH-3); A^3B^1 (LDH-4); A^4B^0 (LDH-5). This hypothesis (Markert, 1962; Appella and Markert, 1961), which owes much to our knowledge of hemoglobin composition and synthesis, not only provides a plausible explanation for the structure of isozymes, but is readily subject to experimental test. At least three tests are apparent. (1) Recombination of equal numbers of the dissociated subunits of LDH-1 and LDH-5 should produce all five isozymes in the ratio of 1:4:6:4:1. The major difficulty in this approach is to discover the conditions required to promote reassociation of the subunits. Some success has been achieved but satisfactory completion of this test will require more experimentation. (2) Total amino acid analyses of LDH-1, LDH-3, and LDH-5 should demonstrate whether LDH-3 has an amino acid composition that could be formed by equal numbers of the subunits found in LDH-1 and LDH-5 as the hypothesis predicts. The results of this test will be presented below. (3) If each type of subunit is antigenic, then immunochemical tests should demonstrate that LDH-1 and LDH-5 are not cross-reactive but that each is cross-reactive with LDH-2, -3, and -4. Such tests have been conducted in several laboratories and the results are consistent with the hypothesis (Plagemann *et al.*, 1960a; Cahn *et al.*, 1962; Lindsay, 1962; Markert and Appella, 1963).

Recombination of dissociated subunits to form new isozymes is the most critical test. Although this test is not yet complete it is interesting to note that most of the patterns of isozymes as seen in zymograms of mouse tissues (Fig. 3) could be produced by a random assortment of subunits, provided that the proportions of the subunits were fixed at an appropriate ratio (Fig. 6). Thus in mouse skeletal muscle the ratio of A to B should be about 30:1, in kidney about 1:10, and in adult heart about 1:3. At about the time of birth the mouse heart should contain equal numbers of subunits A and B, thus producing mostly LDH-3. The few exceptions to the possibility of randomly generated patterns, such as the diaphragm, can be attributed to the heterogeneity of the tissue components, each contributing a quite different isozyme composition to the over-all pattern of the organ.

Analysis of the isozyme patterns in fragments of the stomach (Fig. 3) demonstrates the local heterogeneity that can exist in complex organs or tissues. It is perhaps surprising that the organ patterns agree as closely as they do with patterns predicted upon the basis of random assortment of subunits. The fact that the spectrum of isozymes in any tissue is continuous is also in accord with the subunit hypothesis. No isozymes are skipped in the series and the quantities of each generally follow a smooth gradient of diminution from the most abundant to the least. These characteristics of the

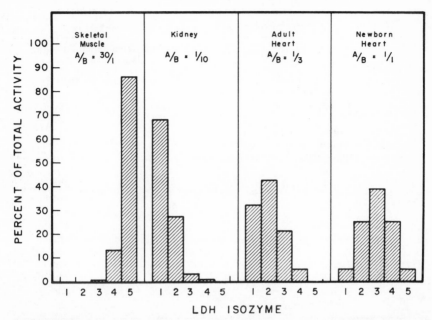

Fɪɢ. 6. Relative abundance of each LDH isozyme in tissues as predicted on the assumption of a random assortment of different initial numbers of the polypeptide subunits A and B.

distribution pattern would be predicted from the subunit hypothesis, provided the subunits were assorted at random.

Chemical Composition

By investigating the chemical composition of isozymes some insight may be obtained into their synthesis and perhaps into their function. Two general approaches are available: determination of total amino acid composition by means of an amino acid analyzer and determination of peptide patterns ("fingerprinting") after trypsin digestion. Both procedures have been applied to crystalline preparations containing several isozymes and also to pure preparations of LDH-1 and LDH-5 from beef tissues. The amino acid composition of these preparations is shown in Table II. It is obvious that LDH-1 and LDH-5 differ considerably although they appear to be related proteins. Moreover the differences correspond with the electrophoretic behavior of these isozymes. At pH 7.0 LDH-1 has a greater net negative charge than LDH-5 and accordingly is richer in aspartic and glutamic acids, but poorer in the basic amino acids, arginine, lysine, and histidine. The basic amino

TABLE II

AMINO ACID COMPOSITION OF LDH ISOZYMES FROM BEEF MUSCLE

Amino acid	Number of amino acid residues per molecule of enzyme[a]	
	LDH-1	LDH-5
Lysine	94	95
Histid ne	25	34
Arginine	34	52
Aspartic acid	123	104
Threonine	56	62
Serine	92	61
Glutamic acid	124	135
Proline	42	63
Glycine	91	100
Alanine	72	122
Valin	135	82
Methionine	32	20
Isoleuc ne	86	73
Leucine	130	118
Tyrosine	26	35
Phenylalanine	19	26

[a] Based upon a molecular weght of 135,000 (including 12 residues of cysteine and 30 residues of tryptophan in each isozyme).

acids are all more abundant in LDH-5. The subunit hypothesis predicts that the amino acid composition of LDH-3 should be equal to one-half the sum of the amino acids of LDH-1 and LDH-5. Unfortunately, pure preparations of LDH-3 in sufficient quantity to provide a critical test of this hypothesis have not yet become available, and preliminary results with small quantities have been ambiguous.

Since the amino acid compositions of LDH-1 and LDH-5 are different, their peptide patterns should also be different, and this is so. Peptides are obtained by digesting the LDH with trypsin, which splits the molecule at each arginine and lysine residue. The resulting peptides are resolved on paper by chromatography and electrophoresis and visualized with the aid of ninhydrin. Such peptide patterns show that some of the peptides from LDH-1 and LDH-5 appear to be the same. The N-terminal residue appears to be threonine in both LDH-1 and -5 and the C-terminal residue is aspartic acid for both isozymes. These similarities argue for a common origin during the biochemical evolution of these isozymes. The two different genes which presumably encode the constituent polypeptides of LDH probably arose by duplication from a single precursor gene and then diverged through the accumulation of spontaneous mutations.

The peptide analysis also tends to confirm the subunit hypothesis. The number of arginine + lysine residues found in LDH-1, for example, is about 130. If only a single long polypeptide were involved, then digestion with trypsin should yield approximately 130 peptides. However, only about one-fourth this number is actually observed (Fig. 7)—a result to be expected

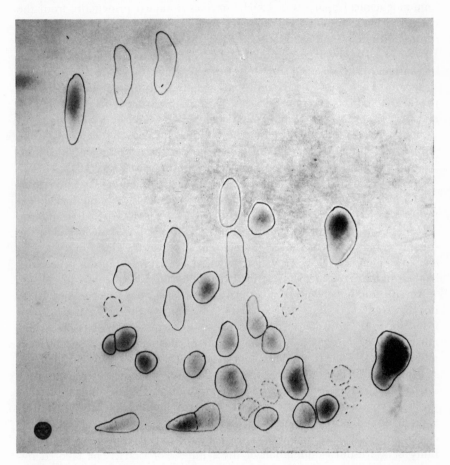

FIG. 7. Peptide pattern of LDH-1 from beef heart. Isozyme digested with trypsin and peptides resolved on paper by electrophoresis and chromatography. The peptide pattern of LDH-5 is conspicuously different. Note that the number of peptides is about 30 or one-fourth of the number of arginine + lysine residues in the intact enzyme molecule.

if the molecule is composed of four identical subunits. It is interesting to note that four molecules of NAD are bound to each molecule of LDH, presumably one per subunit, although once dissociated the subunits do not bind NAD.

Immunochemical Properties

Recently Lindsay (1962) and Cahn *et al.* (1962) have completed extensive immunochemical analyses of LDH isozymes obtained principally from the chicken. These studies show that LDH-1 and LDH-5 are immunochemically distinct although both are related to LDH-2, -3, and -4. Antisera to either LDH-1 or LDH-5 will precipitate LDH-2, -3, and -4 as well as the immunizing antigen, but will not precipitate the isozyme at the opposite end of the spectrum. These results are consistent with the assumption that the polypeptides A and B are distinct antigens. It would then follow that the immunochemical properties of the intact LDH molecule would be a function of its subunit composition. In fact Lindsay's (1962) analysis indicates that one antibody molecule is sufficient to inhibit one enzyme molecule. Thus complete inactivation of LDH-2, -3, and -4 by antisera to either LDH-1 or -5 is to be expected. The data on quantitative precipitin tests by Cahn *et al.* (1962) also supports this expectation. Earlier studies of the immunochemical properties of LDH isozymes of the rabbit (Plagemann *et al.*, 1960a) and frog (Nace *et al.*, 1961) can also be fitted into this general analysis although the subunit hypothesis was not considered in this work. These investigators tended to picture each isozyme as distinct, as indeed it is, but not necessarily in the qualitative sense implicit in their discussions.

Recent immunochemical investigations of beef, pig, and mouse LDH isozymes (Markert and Appella, 1963) also support the conclusions drawn from the work with chick LDH. Isozymes from these different species did not cross-react in agar gel diffusion analyses, but such species specificity is not surprising. During the long course of evolution the genes encoding the structure of LDH in different species would surely become different in some degree. In fact the electrophoretic behavior of the LDH isozymes of different species is commonly different, and the different mobilities indicate a difference in amino acid composition—that is, in the primary structure of the constituent polypeptide chains. This species specificity opens up the possibility of a fruitful examination of the synthesis of LDH. The favored hypothesis of LDH synthesis assumes that two nonallelic genes produce two different polypeptides which, assorting at random in groups of four, produce the five isozymes observed. A hybrid between two species would possess four such genes, each producing a different polypeptide. Assorting four different polypeptides in all possible combinations of four would give rise to 35 different molecular varieties of LDH. Thus an examination of LDH patterns in hybrid tissues should prove very instructive.

Function of Isozymes

From the point of view of cell function it seems strange that a cell should synthesize precise patterns of isozymes, all performing exactly the same catalytic function. One type of molecule would seem sufficient. However, a detailed examination of the catalytic properties of the LDH isozymes shows that they are not identical although each exhibits the basic properties of LDH. The extensive work of Kaplan and his associates (Kaplan *et al.*, 1960; Kaplan and Ciotti, 1961) has clearly demonstrated that the different isozymes of LDH have significantly different catalytic efficiencies, particularly when various NAD analogs are substituted for the normal coenzyme in the assay mixture. This behavior of the isozymes *in vitro* with abnormal substrates probably reflects differences in their normal behavior *in vivo*, although no direct test of this assumption has been made. It seems quite likely that the fundamental role of LDH is to regulate the ratio of NAD to $NADH_2$ with the production of lactate as only an essential by-product of this activity. The $NAD/NADH_2$ ratio is of critical importance in regulating numerous biochemical reactions in the cell. Each isozyme may serve to regulate this ratio at specific locations in the cell where that particular isozyme is bound by virtue of its unique charge.

The catalytic efficiency of the different isozymes has been measured in the presence of different concentrations of the normal substrates, lactate and pyruvate, by several investigators (Plagemann *et al.*, 1960b; Kaplan and Ciotti, 1961; Lindsay, 1962; Cahn *et al.*, 1962; Markert and Ursprung, 1962). These investigations all agree in demonstrating that high pyruvate concentrations inhibit LDH (Fig. 8). However, the substrate optimum for LDH-1 is much lower than for LDH-5, and with increasing concentrations of pyruvate LDH-1 activity is inhibited long before LDH-5 activity is depressed. This behavior has interesting physiological implications and provides a metabolic rationale for the existence of LDH isozymes. Mammalian skeletal muscle, for example, is rich in LDH-5. During periods of vigorous activity this muscle produces large quantities of pyruvate that is then reduced to lactate by LDH-5 with a simultaneous oxidation of $NADH_2$ to NAD. This reaction enables the muscle to use glucose as a source of energy even after oxygen is exhausted, because essential supplies of NAD can be regenerated from $NADH_2$ by transferring the hydrogen to pyruvate (to yield lactate) rather than to oxygen. Thus lactate can serve as a temporary hy-

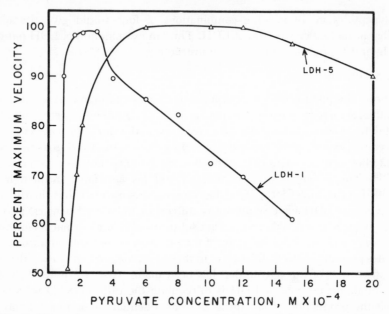

F~IG~. 8. Effect of pyruvate concentration on the velocity of the reaction with human LDH isozymes LDH-1 and LDH-5 at pH 7.0. (Redrawn and modified from the data of Plagemann *et al.*, 1960.)

drogen storage reservoir until increased oxygen supplies become available. However, at high levels of lactate, muscle function is impaired. This is not usually serious in skeletal muscle but it might be fatal in heart muscle. Since heart muscle contains principally LDH-1, large quantities of lactate are not likely to accumulate during rapid metabolism of glucose because the enzyme would be increasingly inhibited by the increased concentration of pyruvate. Thus in heart muscle, lactate production should be maintained at a relatively constant low level even in the face of fluctuating pyruvate concentrations. Only tissues capable of tolerating relatively anaerobic conditions are rich in isozymes at the LDH-5 end of the spectrum. Other tissues are equipped principally with LDH-1 or LDH-2. It is interesting to note that mammalian embryos have a preponderance of LDH-5 in accord with their relatively anaerobic environment. The tissues of chick embryos, on the other hand, appear to be as well oxygenated as adult tissues and these embryonic tissues synthesize mostly LDH-1. Only later in adult tissues subject to anaerobiosis, such as skeletal muscle, does LDH-5 become a prominent part of the enzymatic repertory.

These observations not only give meaning to the existence of isozymes, but suggest that the epigenetic control of their synthesis may in some way involve the oxygen tension in the cell. The metabolic mechanisms which bring about the differential function of genes still remain one of the greatest mysteries in biology, but it does seem clear that regulation of the relative levels of function of two genes for the A and B polypeptides is sufficient to account for the LDH isozymic patterns observed in adult tissues, as well as the changing patterns apparent during embryonic development.

Whether differential gene function is a plausible explanation for the origin of the isozymic forms of enzymes other than LDH, only further investigation will reveal. Although the isozymes of any enzyme carry out the same catalytic function, their distinguishing physical properties may enable them to be integrated into distinct metabolic pathways in different locations in the cell. This arrangement should permit a more precise and sensitive control of cell metabolism and thus should have proved biologically advantageous during the course of evolution. The large number of enzymes which have so far been demonstrated to exist in multiple molecular forms indicates that organisms have frequently exploited the advantages of synthesizing isozymes. These specialized forms of an enzyme must now be accorded a position of general biological significance. Further study of the nature of isozymes and of the genetic and epigenetic mechanisms that regulate their synthesis will surely advance our understanding of the processes of cellular differentiation, and may even bring a solution of the dilemma posed by different phenotypes arising from the same genotype.

Acknowledgments

The work from the author's laboratory reviewed here has been supported by grants from the National Science Foundation, the American Cancer Society, and by contracts with the Atomic Energy Commission. Most of the research was carried out in collaboration with Drs. Ettore Appella and H. Ursprung. The skillful assistance of Joanne Yundt and Sheila Hutman is gratefully acknowledged.

References

Allen, J. M. (1961). Multiple forms of lactic dehydrogenase in tissues of the mouse: Their specificity, cellular localization, and response to altered physiological conditions. *Ann. N. Y. Acad. Sci.* **94,** 937–951.

Appella, E., and Markert, C. L. (1961). Dissociation of lactate dehydrogenase into subunits with guanidine hydrochloride. *Biochem. Biophys. Research Communs.* **6,** 171–176.

Beermann, W. (1961). Ein Balbiani-Ring als Locus einer Speicheldrüsenmutation. *Chromosoma* **12,** 1–25.

Cahn, R. D., Kaplan, N. O., Levine, L., and Zwilling, E. (1962). Nature and development of lactic dehydrogenases. *Science* **136,** 962–969.

Clever, U. (1961). Genaktivitäten in den Riesenchromosomen von *Chironomus tentans* und ihre Beziehungen zur Entwicklung. I. Genaktivierungen durch Ecdyson. *Chromosoma* **12,** 607–675.

GALL, J. G., AND CALLAN, H. G. (1962). H³ Uridine incorporation in lampbrush chromosomes. *Proc. Natl. Acad. Sci. U.S.* **48**, 562–570.

HUNTER, R. I.., AND MARKERT, C. I.. (1957). Histochemical demonstration of enzymes separated by zone electrophoresis in starch gels. *Science* **125**, 1294–1295.

JACOB, F., AND MONOD, J. (1963). Regulation of protein synthesis in bacteria as a model for genetic repression, allosteric inhibition and differentiation. This volume, pp. 30–64.

KAPLAN, N. O., AND CIOTTI, M. M. (1961). Evolution and differentiation of dehydrogenases. *Ann. N. Y. Acad. Sci.* **94**, 701–722.

KAPLAN, N. O., CIOTTI, M. M., HAMOLSKY, M., AND BIEBER, R. E. (1960). Molecular heterogeneity and evolution of enzymes. *Science* **131**, 392–397.

LINDSAY, D. T. (1962). Developmental patterns and immunochemical properties of lactate dehydrogenase isozymes from the chicken. Thesis, Johns Hopkins University, Baltimore, Maryland.

MARKERT, C. L. (1962). Isozymes in kidney development. *In* "Hereditary, Developmental, and Immunologic Aspects of Kidney Disease" (J. Metcoff, ed.), pp. 54–63. Northwestern Univ. Press, Evanston, Illinois.

MARKERT, C. L., AND APPELLA, E. (1961). Physicochemical nature of isozymes. *Ann. N. Y. Acad. Sci.* **94**, 678–690.

MARKERT, C. L., AND APPELLA, E. (1963). Immunochemical properties of lactate dehydrogenase isozymes. *Ann. N. Y. Acad. Sci.* in press.

MARKERT, C. L., AND MØLLER, F. (1959). Multiple forms of enzymes: Tissue, ontogenetic, and species specific patterns. *Proc. Natl. Acad. Sci. U. S.* **45**, 753–763.

MARKERT, C. L., AND URSPRUNG, H. (1962). The ontogeny of isozyme patterns of lactate dehydrogenase in the mouse. *Develop. Biol.* **5**, 363–381.

MEISTER, A. (1950). Reduction of α, γ-diketo and α-keto acids catalyzed by muscle preparations and by crystalline lactic dehydrogenase. *J. Biol. Chem.* **184**, 117–129.

NACE, G. W., SUYAMA, T., AND SMITH, N. (1961). Early development of special proteins. *Symposium on Germ Cells and Development (Inst. Intern. Embryol. and Fondazione A. Baselli), 1960* pp. 564–603.

NEILANDS, J. B. (1952). Studies on lactic dehydrogenase of heart—purity, kinetics, and equilibria. *J. Biol. Chem.* **199**, 373–381.

PFLEIDERER, G., AND JECKEL, D. (1957). Individuelle Milchsäuredehydrogenasen bei verschiedenen Säugetieren. *Biochem. Z.* **329**, 370–380.

PLAGEMANN, P. G., GREGORY, K. F., AND WRÓBLEWSKI, F. (1960a). The electrophoretically distinct forms of mammalian lactic dehydrogenase. I. Distribution of lactic dehydrogenases in rabbit and human tissues. *J. Biol. Chem.* **235**, 2282–2287.

PLAGEMANN, P. G., GREGORY, K. F., AND WRÓBLEWSKI, F. (1960b). The electrophoretically distinct forms of mammalian lactic dehydrogenase. II. Properties and interrelationships of rabbit and human lactic dehydrogenase isozymes. *J. Biol. Chem.* **235**, 2288–2293.

SMITHIES, O. (1955). Zone electrophoresis in starch gels: Group variations in the serum proteins of normal human adults. *Biochem. J.* **61**, 629–641.

WALKER, D. G., AND SELIGMAN, A. M. (1963). The use of formalin fixation in the cytochemical demonstration of DPN and TPN dependent dehydrogenases in mitochondria. *J. Cell Biol.* in press.

WIELAND, T., AND PFLEIDERER, G. (1957). Nachweis der Heterogenität von Milchsäuredehydrogenasen verschiedenen Ursprungs durch Trägerelektrophorese. *Biochem. Z.* **329**, 112–116.

Quantitative Studies of Protein Synthesis in Some Embryonic Tissues

HEINZ HERRMANN

Institute of Cellular Biology, University of Connecticut, Storrs, Connecticut

Each torpid turn of time has those disinherited children
To whom no longer what's been, and not yet what's coming, belongs.

Rilke

During the last decade, knowledge of the chemical reactions leading to synthesis of proteins has increased at an unprecedented rate. Some of the main lines of research which are beginning to unfold a coherent picture of one of the key processes of cell life may be mentioned here. Included are the investigations of the mechanisms for amino acid transfer into the cells and their concentration in the free amino acid pool (Christensen *et al.*, 1948; Cohen and Rickenberg, 1956; Piez and Eagle, 1958; Cowie and McClure, 1959; Christensen, 1959). Prominent advances have been made in the exploration of the reactions involved in the activation of amino acids leading to the formation of an amino acid adenylate in the presence of adenosine triphosphate (ATP), the transfer of the amino acid first to a soluble ribonucleic acid (RNA) fraction and then to the ribosomal protein. An analysis has been carried out of the structure and synthesis of the terminal group of the transfer RNA and of the enzymes which catalyze these transfer reactions (Work, 1959; Zamecnik, 1960, 1962; Hoagland, 1960; Nathans *et al.*, 1962). A major effort has been directed toward the exploration of the role of ribosomes as the main site of protein synthesis in the structural organization of the cell; there the main advances rest on complementary work with biochemical analysis and analysis of fine structure (Palade, 1958; Roberts, 1958). Evidence has been obtained for protein synthesis in the cell nucleus (Allfrey *et al.*, 1960; Sibatani *et al.*, 1962), and for the transfer of nuclear RNA for cytoplasmic protein synthesis (Prescott, 1962; McMaster-Kaye, 1962; Leblond and Amano, 1962).

Decisive advances in the understanding of the cellular control of protein synthesis have been made by the studies extending over the last decade of enzyme adaptation in microbial forms (Cohn, 1957; Pollock, 1959) and of enzyme repression and feedback inhibition in bacteria (Pardee, 1962). This work has led recently to new concepts of genic control of protein synthesis and to experimental evidence for gene products, such as messenger RNA, which are involved in the control of the protein-forming system (Frisch, 1961; Jacob and Monod, 1961; Nirenberg and Matthaei, 1961; Volkin, 1962; Ingram, 1962).

The pertinence of all these new discoveries and concepts for a reinterpretation of embryonic growth and development is quite apparent. It seems self-evident that the changes in the rates of protein formation during embryonic development, and the appearance and disappearance of enzymes and of other cell constituents which secondarily lead to cell diversification and to morphogenesis, must be accompanied by changes in the quantities and qualities of the components of the protein-forming system and of its controlling mechanism. In turn, it might be expected that the application of the new treasure of information to the study of embryonic tissues would provide the investigator of protein synthesis with valuable opportunities to check his theories and model systems against the hard realities of embryonic development.

Desirable as it is, very little interaction has occurred between the fields of developmental biology and experimental embryology, on the one hand, and that of protein synthesis, on the other hand. Until a few years ago (Ebert, 1955; Herrmann, 1960), the disparity in the understanding of protein synthesis in model systems and in embryonic tissues did not seem so formidable. Since then, the gap between the two fields seems to have widened, largely because of the accelerated pace of research on protein synthesis. In attempting now to close this gap, must one give up the safe harbor of experimental embryology in order to reach the new continent of protein biosynthesis in development? It is this question which elicits the motto used for the introduction to this paper. At the same time, it is hoped that the following discussion will justify a more affirmative paraphrasing of these words in the concluding lines of this article.

Before turning now to the explorations of protein synthesis in embryonic tissue carried out at our laboratory, and to the work of other investigators which is related to it, other pertinent approaches to the same subject should be mentioned.

No reference will be made, at this point, to work dealing solely with the accumulation of proteins during development. An important aspect of this subject has been discussed in the preceding paper (see chapter by Markert) of this volume and a more comprehensive review, covering this topic in more detail, will be published elsewhere (Herrmann and Tootle, in prep.). The work emphasized is that attempting to establish a closer connection between protein accumulation and the activity of the protein-forming system of the developing cell. The analysis of the changes in the incorporation of amino acids into subcellular fractions of the sea urchin egg following fertilization was initiated by Hultin (1951, 1953a,b). More recently investigators at the University of Palermo have measured the transfer of labeled amino acids from the amino acid pool to various soluble and particulate fractions after fertilization (Nakano and Monroy, 1958) of the sea urchin egg. It could be shown that transfer of label into soluble protein (Monroy *et al.*, 1961) and into microsomes (Monroy, 1960) ensues a few minutes after fertilization, but the uptake into mitochondria occurs only after several hours (Giudice and Monroy, 1958). The rate of uptake into the proteins of the sea urchin embryo was found to reach a maximum at the 32–64-cell stage (Giudice *et al.*, 1962). Hultin and Bergstrand (1960) and Hultin (1961) suggest some change in the state of the ribosomes after fertilization as an explanation of the absence of ribosomal incorporation before fertilization and the rapid increase of incorporation after fertilization. In ascidian embryos, uptake of amino acids into proteins begins before fertilization (Reverberi *et al.*, 1960).

In embryonic tissues of *Xenopus laevis*, characteristic differences in amino acid activation (Deuchar, 1961, 1962) and incorporation (Deuchar, 1960) were observed. A high activation rate was found during the neurula stage, and among the embryonic organ primordia the somites showed a maximal activation of leucine. The uptake of labeled amino acids into early amphibian and chick embryos was followed autoradiographically by Waddington and Sirlin (1954) and Feldman and Waddington (1955). Main sites of the intracellular protein synthesis in embryonic tissues were suggested by Sirlin and Waddington (1956) and their changes in amino acid incorporation during development of mesodermal tissue were reported by Waddington and Sirlin (1959).

Changes in the protein-forming system, implicating the role of RNA, during the establishment of hemoglobin synthesis have been described in the promising approaches by O'Brien (1961) and Wilt (1962). As an example of a study of the protein-forming systems during later organogenesis, recent work by Flexner *et al.* (1958) on the utilization of C^{14}-glucose for synthesis of amino acids and proteins and their metabolic interconversion in the liver and brain of newly born mice should be mentioned.

Evidence for enzyme induction in the development of tryptophan pyrrolase has been negated by Spiegel and Frankel (1961), but interesting experiments supporting the occurrence of enzyme repression during development of transamidinase were carried out by Walker and Walker (1962).*

* While this paper was in press significant contributions have appeared on the initial development of ribosomes (Brown and Caston, *Develop. Biol.* 5, 412, 1962) and synthesis of large molecular RNA (Brown and Caston, *ibid.* 5, 435, 1962) in amphibian

In our laboratory, three systems have been selected for the study of protein biosynthesis in development: (1) the explanted chick embryo, (2) differentiation of the corneal and scleral mesenchyme, and (3) development of the leg musculature. Each of these systems seems to offer distinct advantages for the exploration of the relation of protein synthesis to embryonic development. In the explanted chick embryo, analysis of protein synthesis and breakdown is possible under conditions such that development proceeds quite similarly to that in the intact embryo. Yet the opportunity to change experimentally the variables of this system and the analyzability of the processes connected with maintenance of protein synthesis are much greater than in the embryo *in ovo*. The mesenchyme of the eye is a tissue with a very uniform cell population in which cell diverisfication occurs at a relatively late date of development and can be readily defined on the molecular level in quantitative and qualitative terms. The relation between cell proliferation and the production of specific cell proteins can be readily studied and *in vitro* synthesis can be measured because of the unusual viability of these tissues outside the body. The third system is the developing muscle in which the components of the protein-forming system during different stages of development can be characterized and the observed changes related to the distinctive phases in the development of this tissue, in particular to the formation of the proteins of the myosin group.

Compared with the more advanced studies on model systems of protein synthesis, the work reported here is at best an elementary quantitation. It is felt, however, there may be some merit in surveying the work now done with several embryonic systems and to learn what analysis can be carried out in each of them.

I. Protein Synthesis in the Explanted Chick Embryo

A. *The Reproducibility* in Vitro *of Growth and Development* in Vivo

The chick embryo, as many other forms of avian embryos, can be more readily separated from the bulk of the yolk than the amphibian embryos with their large intracellular store of yolk, and it is, of course, much less dependent upon the maternal organism than the embryos of mammals. This particular position of the avian embryo has tempted investigators for some time to

---

embryos, on the interrelation of messenger polynucleotides and amino acid incorporation in developing sea urchin eggs (Nemer, *Biochem. Biophys. Research Communs.* **8,** 511, 1962; Wilt and Hultin, *ibid.* **9,** 313, 1962), the effect of actinomycin on development, and protein and RNA synthesis in sea urchin embryos (Gross and Cousineau, *Abstr. 2nd Meeting, Am. Soc. Cell Biol.,* 1962), and the effect of fluorouracil on development of tryptophan pyrrolase (Nemeth, J. Biol. Chem. **237,** 3703, 1962).

study factors which are essential for its maintenance, growth, and development outside of the egg. A technique for the explantation of avian embryos was first employed by Waddington (1932) and later elaborated by Spratt (1947). Morphogenesis in the explanted avian embryo can be maintained to a far-reaching extent under the tested conditions. The usefulness of this system for experimental embryology has been indicated by the studies of Spratt (1958), McKenzie and Ebert (1960), and at our laboratory (Herrmann, 1953; Rothfels, 1954; Hermann *et al.*, 1955; Schultz and Herrmann, 1958; Schultz, 1959). Characteristic responses of these explants to metabolic inhibitors and amino acid analogs were demonstrated. Of interest also is the use of these explants for the production and analysis of abnormal forms of development, resembling chemically induced phenocopies in chick embryos developing in the egg.

The quantitative aspects of protein accumulation in explanted chick embyros during the explantation for periods from 8 hours to 24 hours and more were investigated during the past years at this laboratory (Britt and Herrmann, 1959; Herrmann and Schultz, 1958; Hayashi and Herrmann, 1959; Deuchar and Herrmann, 1962). Systematic efforts to develop conditions under which growth and development of the explanted chick embryo occur optimally *in vitro* are being pursued in our laboratory by Klein *et al.* (1962a). A comparison of the minimal and optimal rates of increase in protein content in which embryo explants with the protein content of the chick embryos *in ovo* is given in Fig. 1. It can be seen that on synthetic medium containing no amino acids (basal medium) and with air and CO_2 as gas phase, an almost negligible protein increase of the embryo occurs, while the formation of somites proceeds at an almost undiminished rate. Thus, certain phases of differentiation progress normally even when net protein accumulation is almost completely suppressed. Under optimal growth conditions with a medium containing whole egg constituents as nutrient mixture and a carefully graded oxygen level in the gas phase, accumulation in the explants approaches closely that of *in ovo* during the first 24-hour period. It drops to a rate of about 65% of the *in ovo* control on the second day and eventually falls to about one-half of the control rate during the third day of *in vitro* culture (Klein *et al.*, 1962b). These data show that during the first day of explantation growth and development *in vitro* can quite closely duplicate that which occurs *in ovo*, and that during the second and third day of explantation a substantial part of the protein increase of the *in ovo* embryo can be maintained in culture. Under conditions of both minimal and maximal growth, normal formation of many organ primordia occurs. Thus, an independence of the rates of growth and of morphogenesis is demonstrated.

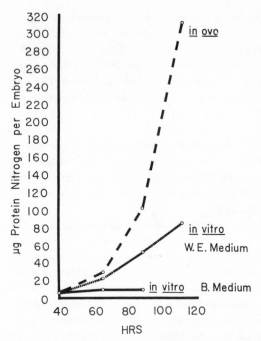

FIG. 1. Increase in protein nitrogen of chick embryos (axial structures) *in ovo* and *in vitro* for the period of development 40–112 hours. W. E. Medium: whole egg homogenate agar medium. Increasing O_2-5% CO_2 as gas phase. B. Medium: basal medium. Electrolyte, glucose, vitamin, agar medium without amino acids or protein. The embryos were explanted at 40 hours of development *in ovo* for subsequent explantation periods of 24, 48, and 72 hours. This corresponds to the total time of incubation, indicated on the abscissa, of 64, 88, and 112 hours.

This dissociation of growth and development has been commented upon on previous occasions (Needham, 1942; Stockdale and Holtzer, 1961).

These improvements in explantation technique provide opportunities for several forms of analysis of the explants. It is now possible to investigate the relation of amino acid incorporation to different rates of net protein increase. Whether proteins of the different organ primordia and individual proteins are proportionately or disproportionately affected during development at widely varying rates of protein increase may be studied. The significance of the role of protein degradation during growth may be demonstrated. In offering such advantages in analyzability, the explant can be used as an experimental model for exploration of the mechanisms of development of the embryo *in ovo*. At the same time, data obtained on explants, in which most of the natural relations between organ primordia exist, can serve as base-lines for comparison with a variety of chemical growth parameters in isolated embryonic tissues growing in organ culture and even for dispersed or reag-gregated cell cultures which are obtained from the tissues of the embryo. Some of the analyses carried out on such explants are described in the follow-ing section.

B. *The Relation of Protein Increase (Growth) to Amino Acid Incorporation and Release in Explant Proteins*

As a first step in attempting to relate the protein increase of the chick embryo explants to activity of the protein-forming system, comparisons were carried out of the rates of protein increase to the rates of amino acid incorpo-ration into the proteins of the explants. In the simplest case, there should be a direct proportionality in protein increase and amino acid incorporation under conditions giving widely differing rates of protein increase. However, this expectation is fulfilled only if two assumptions are met. There must be no change in the size of the free amino acid pool in the cells of the explants, and the rate of protein degradation either must be negligible or its magnitude must remain a constant proportion of the total protein accumulation. If the amino acid pools vary, the added radioactive amino acid is diluted to differ-ent extents and no strict correlation of tracer incorporation and net protein increase can be expected.

These considerations became important in comparing growth and incor-poration of explants which were maintained either on whole egg medium or on a synthetic medium (Hayashi and Herrmann, 1959). Growth on whole egg medium led to a rapid protein increase with minimal tracer uptake, while on certain synthetic media very small net protein increases were ob-tained together with a relatively large incorporation of tracer amino acids. Exactly where the dilution of tracer took place when the explants were grown on whole egg medium is still under investigation. However, this example shows that tracer dilution in the cell is an important variable in relating pro-tein increase and protein synthesis.

Since amino acid pool size can be quantitatively measured, the variability of this magnitude does not pose an unsurmountable obstacle for a comparison of tracer incorporation and protein increase in explants, in which widely differing rates of protein increase are brought about by cultivation on differ-ent nutrient media. However, in order to simplify these experimental condi-tions in the beginning of such an analysis, we turned to explantation of the chick embryo on media containing amino acid analogs. In the presence of these substances we expected a direct relationship between amino acid in-

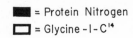

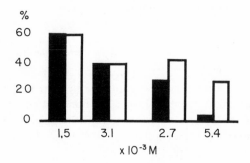

Chloramphenicol   Fluorophenylalanine

FIG. 2. Effects of fluorophenylalanine and chloramphenicol on protein nitrogen content and incorporation of glycine-1-C¹⁴ of chick embryos after explantation for 24 hours on whole egg agar medium. The figures on the abscissa indicate final concentration of analogs in medium. The figures on the ordinate indicate the values for the embryos growing on analog-containing medium in per cent of the values for control embryos explanted on medium without analog (100%).

corporation and net protein increase under widely differing rates of protein increase. Contrary to this expectation, when the effects of two analogs of phenylalanine were tested, different results were obtained (Fig. 2). With chloramphenicol, the increases in protein nitrogen and tracer incorporation were inhibited to the same extent. With fluorophenylalanine, the increase in protein nitrogen was significantly more reduced than the tracer incorporation. To account for this discrepancy, it was assumed that some protein breakdown occurred in the samples grown in the presence of fluorophenylalanine or that the amino acid pool was decreased on addition of this analog. From the work of Cohen *et al.* (1958) it has become apparent that the amino acid pool in yeast cells is not diminished, and fluorophenylalanine is incorporated into cellular proteins, but these analog-containing proteins are broken down more rapidly than the normal proteins. Hancock (1960) found also an increased amino acid pool after inhibition of protein synthesis by fluorophenylalanine and chloramphenicol. By addition to the explantation medium of C¹⁴-labeled fluorophenylalanine at concentrations below those necessary for inhibition of amino acid incorporation, it was demonstrated that appreciable amounts of the analog were incorporated into the protein

TABLE I

Incorporation of Phenylalanine and Fluorophenylalanine into Chick Embryo
Explant Proteins During a 24-Hour Explantation Period[a,b]

	Protein nitrogen (μg)	Total activity (counts/min)	Specific activity (counts/min/mg protein)
Phenylalanine[c]	24.7	3,850	157.1
Fluorophenylalanine[d]	21.8	1,961	90.2

[a] Adapted from Herrmann and Marchok (1963).
[b] Explantation conditions: large trim, whole egg medium.
[c] Specific activity: 1.3 μC/millimole; 1 μC/ml.
[d] Specific activity: 3.5 μC/millimole; 1 μC/ml.

of the explant (Table I). The analog is incorporated at about one-sixth of the rate of phenylalanine, if the results are corrected for differences in specific activities of the compounds used. Therefore, it seems likely that the discrepancy between protein increase and amino acid incorporation in explants grown in the presence of fluorophenylalanine is due to a degradation of proteins as in the microbial cell, and that the more rapid degradation may be attributed to formation of abnormal analog-containing proteins. In the presence of chloramphenicol, which is not incorporated into proteins, no discrepancy between protein increase and amino acid incorporation was observed. Therefore, in the presence of chloramphenicol, no evidence for an increased breakdown of proteins could be obtained. The role of chloramphenicol as an inhibitor of protein synthesis is not fully clarified and its analysis is complicated by side effects (Gale, 1958; Rendi, 1959; Dagley and Sykes, 1960; Fraenkel and Neidhardt, 1961).

It should be mentioned at this point that with analogs of leucine used in short-term (8-hour) explants, no inhibition of amino acid incorporation was observed, but the increase in protein content of certain organ primordia was markedly reduced (Schultz and Herrmann, 1958). In the latter case it was assumed that protein synthesis remained unimpaired and that the inhibition of protein increase was entirely due to an acceleration of protein breakdown.

The apparent role of protein degradation in determining net protein increase in chick embryo explants led us to experimentation to demonstrate more directly the existence of protein breakdown under different culture conditions. The explants were grown for a period of 4 hours on a medium containing glycine-1-C^{14}. During this time, well-measurable quantities of the amino acid were incorporated into the proteins of the explants. The excess free glycine-1-C^{14} was removed by washing the explants and by transferring

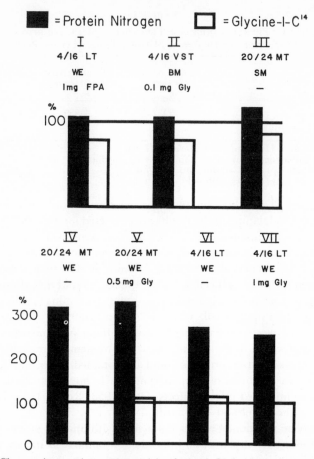

FIG. 3. Changes in protein content and in glycine-1-C[14] level of explants during second explantation period after transfer from medium containing glycine-1-C[14] to medium containing glycine-1-C[12]. The values found at the beginning of the second explantation period on "cold" medium are set equal to 100%.

Conditions of explantation: All explants were kept on the indicated medium with glycine-1-C[14] for 4 hours (first explantation period) followed by two explantation periods of indicated duration (second and third explantation periods) on medium with glycine-1-C[12].

I. Second explantation period 4 hours, third explantation period 16 hours. Embryos explanted with large membrane area (LT) on whole egg homogenate agar medium (WE) containing 1 mg fluorophenylalanine/ml.

II. Second explantation period 4 hours, third explantation period 16 hours. Embryos explanted with very small membrane area (VST) on medium (BM) containing electrolytes, glucose, vitamins, but no amino acids except 0.1 mg glycine/ml.

them to a cold medium for an additional explantation period. At the end of this period, the protein nitrogen and amount of tracer amino acid in the explants were measured in one set of embryos designated as controls. The experimental explants were transferred to another cold medium for varying incubation periods. These experimental embryos were analyzed for protein content and concentration of radioactive amino acid. Thus, it was shown (Fig. 3) that under conditions of rapid growth no decrease of tracer amino acid from the protein of embryos could be observed. However, under conditions of minimal growth, the radioactivity in the explant proteins was diminished although the embryos showed no sign of degeneration on microscopic inspection. As much as 20% of the initial activity was lost during a 24-hour explantation period. This was the case irrespective of whether slow growth was brought about by the presence of amino acid analogs or by growth of explants on a poor nutrient medium (Marchok and Herrmann, 1963).

C. *The Inhibition of Protein Increase in Individual Organ Primordia*

The data discussed so far deal with amino acid incorporation and protein increase in the whole explanted embryo. As the next step in this analysis, it was to be ascertained whether in explants growing at widely differing rates of protein increase and amino acid incorporation, these parameters are affected proportionately in different organ primordia. No systematic study of this question has been carried out as yet, but a distinct independence of the measured parameters could be observed in one series of experiments

---

Legend for Fig. 3 continued

III. Second explantation period 20 hours, third explantation period 24 hours. Embryos explanted with medium membrane area (MT) on medium (SM) containing electrolytes, glucose, vitamins, and amino acids.

IV. Second explantation period 20 hours, third explantation period 24 hours. Embryos explanted with medium membrane area on whole egg homogenate agar medium.

V. Second explantation period 20 hours, third explantation period 24 hours. Embryos explanted with medium membrane area on whole egg homogenate agar medium containing 0.5 mg glycine/ml.

VI. Second explantation period 4 hours, third explantation period 16 hours. Embryos explanted with large membrane area on whole egg homogenate agar medium.

VII. Second explantation period 4 hours, third explantation period 16 hours. Embryos explanted with large membrane area on whole egg homogenate agar medium containing 1 mg glycine/ml.

Conditions I–III lead to little increase of protein nitrogen per embryo and a distinct loss of label from embryos (glycine-1-C^{14} below 100%). Conditions IV–VII lead to a marked increase in protein nitrogen per embryo and no loss of label from embryos (glycine-1-C^{14} equal to or above 100%). Adapted from Marchok and Herrmann (1963).

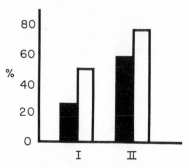

FIG. 4. Effect of 0.83 × 10³ M acetylpyridine on protein nitrogen (solid rectangles) and glycine-1-C¹⁴ (rimmed rectangles) uptake into proteins of nervous system and trunk tissue of chick embryos explanted for 24 hours on whole egg homogenate agar medium. I = central nervous system; II = trunk tissue. Values obtained for controls grown under the same conditions without acetylpyridine are set equal to 100%.

(Buckingham and Herrmann, in prep). The chick embryo explants were grown on media containing 0.8 × 10⁻³ M 3-acetylpyridine, an analog of pyridine nucleotides. The graphs in Fig. 4 show that in the presence of the analog the protein increase of the central nervous system dissected after explantation for 24 hours is reduced to about 26% of the value for the controls and the amino acid incorporation to 49% of the controls. Corresponding inhibitions for the remaining trunk tissue are 56% for the protein increase and 76% for the amino acid incorporation. Hence, both protein increase, as well as amino acid incorporation, are inhibited more strongly in the central nervous system than in the remaining tissues of the embryo. In addition, inhibition of the protein increase is disproportionately greater than the inhibition of amino acid incorporation in both tissues, this discrepancy being larger in the central nervous system than in the remaining trunk tissues.

D. Changes in Individual Protein Content

Finally, in the analysis of protein formation both *in vitro* and *in vivo*, it is found that not all proteins are affected proportionately by variations in the over-all rate of protein accumulation. In a series of experiments (Newburgh *et al.*, 1962) chick embryos were explanted for a 24-hour period on media on which the net protein increase varied widely. Analysis of the explants grown under these different conditions revealed two major experimental groups, one with a high (2.2) protein content per DNA (per cell), and another with a low ratio (1.5) of protein nitrogen to DNA. It was assumed that in the latter group less protein is formed per cell than in the former group. If all proteins followed the same pattern of net increase per cell, the same difference in protein per DNA ratio should hold for each individual protein. Determinations carried out under the two sets of conditions for isocitric dehydrogenase and glucose-6-phosphate dehydrogenase showed that this is not the case. As indicated in Table II, enzymatic activity of isocitric dehydrogenase per DNA remains constant under conditions under which the ratio of total protein to DNA changes from 2.2 to 1.5. It can be expected that this discrepancy will become even greater in determinations of individual proteins in separate organ primordia of the embryo.

E. *Discussion*

By using explants as model systems closely related to the *in ovo* embryo, it was possible to define net protein increase in terms of amino acid incorporation and release, and thus as a function of the protein-forming system. To obtain direct information about the components of the protein-forming system in the early embryo explants poses at the present time formidable methodological difficulties. However, the feasibility of testing the measurable parameters of protein synthesis under a wide variety of experimental conditions, and thus obtaining indirectly some insight into the function of the protein-forming system, has been demonstrated. The use of recently discovered specific inhibitors of protein synthesis and of related synthetic pathways, like puromycin (Morris *et al.*, 1962; Rabinovitz and Fisher, 1962; Gorski *et al.*, 1961) or actinomycin (Kersten and Kersten, 1962; Goldberg and Rabinowitz, 1962), promises further advances in the understanding of protein synthesis even in the early chick embryo explant. Already explants have proved to be among the protein-synthesizing systems which are inhibited by chloramphenicol (Breitman and Webster, 1958; Rabson and Novelli, 1960; Matthaei and Nirenberg, 1961; Nirenberg and Matthaei, 1961) in contrast to others where chloramphenicol was found to be without effect (von Ehrenstein and Lipmann, 1961). Similar to many other bacterial (Cowie *et al.*, 1959; Munier and Cohen, 1959; Kempner and Cowie, 1960) and animal cells (Vaughan and Steinberg, 1958, 1960; Westhead and Boyer, 1961), explants incorporate fluorophenylalanine into proteins. In a certain concentration range, fluorophenylalanine inhibits amino acid incorporation, and at the same time it increases the release of amino acids, presumably due to protein breakdown. Thus, the inhibition of protein net increase is disproportionately larger than inhibition of amino acid incorporation. This is in accordance with data from microbial literature, but it remains undecided whether the increased protein breakdown stems only from the degradation of fluorophenylalanine-containing protein molecules or results in part from increased proteolysis of normal proteins as well. The latter alternative has

TABLE II

Protein Nitrogen, DNA, and Isocitric Dehydrogenase Activity (ICD)
in Explanted and *in Ovo* Chick Embryos[a]

Conditions	Protein N per embryo (µg)	DNA per embryo (µg)	ICD (Micromoles TPN reduced per min per embryo $\times 10^{-4}$)	PN/DNA	ICD (Micromoles TPN reduced per min per DNA $\times 10^{-4}$)
Group I					
40 hr *in ovo* [11–13 somites (49)[b]]	4.1±0.31[c]	2.0±0.29	4.9±0.53	2.05	2.5
64 hr *in ovo* [28–30 somites (32)]	29±2.1	13±1	29±2.9	2.2	2.2
40 hr *in ovo* + 24-hr explantation					
CEH[d]—large trim (56)	17±0.95	7.7±0.88	16±0.45	2.2	2.2
GS[d]—large trim (28)	14±1.3	6.8±0.68	14±2.5	2.04	1.8
Average				2.1	2.2
Group II					
CEH[d]—large trim ± 0.1 mg acetylpyridine/ml (21)	11±0.79	8.5±0.3	13±2.2	1.4	1.5
CEH[d]—small trim (56)	15±1.5	10±0.46	21±2.9	1.5	2.1
CEH[d]—small trim ± 0.1 mg acetylpyridine/ml (42)	7.6±1.3	5.0±0.5	14±1.3	1.5	2.9
88 hr *in ovo* (12)	110	83	180	1.3	2.2
40 hr *in ovo* + 48-hr explantation					
CEH[d]—small trim (15)	20±1.5	16±2.1	42±34	1.2	2.7
Average				1.4	2.3

[a] Adapted from Newburgh, *et al.* (1962).
[b] Number of embryos used.
[c] Standard error.
[d] CEH: concentrated egg homogenate medium. GS: glucose-saline medium.

been tentatively suggested as being more compatible with the data obtained so far (Herrmann and Marchok, 1963).

The results obtained with acetylpyridine may have relevance in several respects. They show that acetylpyridine inhibits protein increase and amino acid incorporation disproportionately to each other and differentially in individual organ primordia. If, as in the case of fluorophenylalanine, disproportionality between protein increase and amino acid incorporation indicates the extent of protein breakdown, acetylpyridine, too, has to be considered to elicit enhanced proteolysis. The mechanism of proteolysis is not well understood at the present time. Possible roles of catheptic activity have been discussed recently (Herrmann and Marchok, 1963; von Hahn and Herrmann, 1962), but mechanisms of energy-linked protein degradation have also been considered (Steinberg and Vaughan, 1956a,b; Halvorson 1958a,b).

Acetylpyridine and other analogs of pyridine nucleotides have frequently been used as teratogenic agents (Ackermann and Taylor, 1948; Landauer, 1957, 1960) in chick embryos. It will be of interest to see whether the teratologically most sensitive organ primordia are distinguished by a particularly high inhibition of both protein increase and amino acid incorporation or by a marked discrepancy of the two parameters, indicating enhanced protein breakdown.

From the data obtained so far, it would seem that protein breakdown, perhaps associated with cell degeneration in individual tissues, plays a minor role in the control of protein increase in the whole embryo growing at the normal rate (Herrmann and Marchok, 1963). However, our data are not as yet conclusive since it is possible, though unlikely, that in rapidly growing explants the loss of label from the embryo is replaced by label from the embryonic membranes. It should be pointed out that, in bacterial growth, protein breakdown and turnover are found (Rickenberg and Lester, 1955; Halvorson 1958a,b; Mandelstam and Halvorson, 1960) to be much greater during the resting phase than during rapid cell proliferation. Little evidence for rapid protein turnover was found in proliferating cells of the intestinal epithelium (Lipkin et al. 1961). Turnover of proteins even in rapidly growing tissue culture cells was demonstrated by Eagle et al. (1959).

II. Protein Synthesis in Differentiating Tissues during Late Development

A. Development of the Connective Tissue of the Cornea and Sclera of the Chick Eye

For the study of protein synthesis in the later phases of development, tissues were chosen which begin to produce large quantities of a single, or a few, well-defined proteins during early periods of organogenesis. The connective tissue of the sclera and cornea of the eye and the leg musculature seemed to be suitable from this point of view. In the fibroblast of the eye envelope one encounters cell types in which, still during embryonic development, more than half of the total protein-producing capacity becomes committed to the formation of a single protein type, namely, collagen. In such tissues, the quantitative relationship of production of specific proteins to production of more generally occurring cell proteins can be investigated in the course of development.

The fibroblasts of the sclera and of the corneal stroma are derived from the head mesenchyme. Both types of connective tissue are anatomically clearly demarcated and can be readily and quantitatively dissected. They consist of layers of apparently homogeneous cell populations except for a few macrophages and nerve axons, and a scanty vascularization in the connective tissue of the sclera which is totally absent in the corneal stroma. The relative avascularity of these tissues and the necessary dependence on diffusion of nutrients *in situ* may be one of the reasons for their high viability *in vitro*. These tissue cell layers are so thin that they allow rapid diffusion of oxygen and nutrients *in vitro* and can thus maintain synthetic activities for many hours *in vitro*. For example, incorporation of amino acids into corneal proteins continues for 6 hours without decline in its rate. In fact, the rate diminishes only by about 30% after 12 hours.

Although derived from the same head mesenchyme, the cells of the sclera and the cornea show a different pattern of chemical diversification during their development. The sclera forms an opaque cartilage containing much more chondroitinsulfate than keratosulfate, whereas in the corneal matrix keratosulfate predominates (Meyer and Chaffee, 1940). As pointed out previously (Herrmann, 1958), the rates of formation of collagen and DNA are quite comparable in the sclera and in the cornea. However, the rate at which the non-collagen proteins are formed differs in the two tissues (Fig. 5). The rates of increase of collagen per DNA begin at a low level at the 8th day of development in both tissues, reach a maximum on about the 16th day, and gradually decline thereafter. The non-collagen proteins in the cornea show an increase per DNA only during the 14–16-day period of development, while the non-collagen proteins in the sclera maintain a rapid rate of increase from the tenth day until the 22nd day of development. The mechanism for these divergent trends in the development of the sclera and cornea is not known. The role of the corneal epithelium in bringing about this divergence is now under study in our laboratory. It must be pointed out here that the apparently marked effect of the epithelium on protein synthesis in

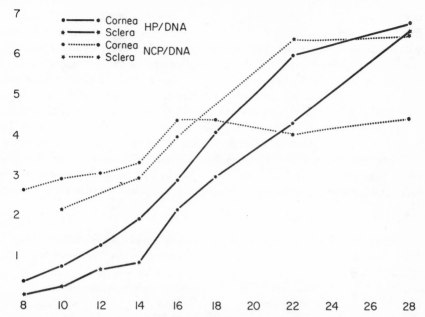

F1G. 5. Comparison of collagen hydroxyproline (HP) content per DNA (HP/DNA) and non-collagen-protein nitrogen (NCP) content per DNA (NCP/DNA) in the corneal stroma and in the sclera of the eye of the developing chick embryo. (From Herrmann, 1958).

the stroma reported previously (Herrmann, 1957, 1961) has been found to be an artifact due to injury of the stroma cells in the course of the separation of the epithelium. The protein synthesis in the stroma can be maintained if the removal of the epithelium is carried out with gentler methods (Herrmann and LeBeau, 1962).

It is of interest that in the cornea the increase of non-collagen proteins, as well as the increase in DNA, comes to a standstill on the 16th to 18th day of development, while collagen synthesis continues for a much longer time. These data represent clearly the inverse relationship between protein synthesis leading to an increase in cell number and protein synthesis leading to an increase in specific cell products, such as collagen. In this sense the data provide an example which shows in quantitative terms the well-known complementarity of cell proliferation and formation of specific cell products (Needham, 1942; Stockdale and Holtzer, 1961). It is likely that the gradual increase in the rate of collagen production is due to the increase in the number of cells in which DNA synthesis with cell proliferation is terminated and protein formation becomes restricted to the synthesis of specific cell products like collagen. This interpretation is supported by findings on other cell types which will be referred to later in more detail. It cannot be rigorously excluded here that the cells of the corneal stroma undergo a more gradual change from the proliferative stage to a state of production of specific cell constituents. What seems important for the study of mechanisms of protein synthesis during embryonic development is the fact that a transition occurs from synthesis connected with cell proliferation to protein synthesis with the predominant formation of a few specialized protein types. The question is whether this transition is reflected in any changes in the properties of the protein-forming system itself. This question will be further explored in the following section.

B. *Protein Synthesis and the Protein-Forming System in Developing Muscle Tissue*

Measurements of protein accumulation, in particular of the proteins of the contractile apparatus of muscle tissue, were initiated for the later stages of the development of rat and chick embryos, using preparative and viscosimetric methods for the assay of this group of proteins (Herrmann and Nicholas, 1948a,b; Csapo and Herrmann, 1951). The early phases of myosin formation have been analyzed immunologically by Ebert (1953), Holtzer *et al.* (1957), Holtzer (1961), Winnick and Goldwasser (1961), and Ogawa (1962), and enzymatic assays have been reviewed by Nass (1962). The main reasons for the choice of muscle as a favorable tissue for the analysis of protein formation in embryonic development has been stated previously (Herrmann, 1952). Muscle is quantitatively the largest single tissue component of most warm-blooded animals. Very soon during maturation, specific cell proteins, the proteins of the actomyosin group, form a major proportion of the total protein moiety. These proteins can be sharply defined and fractionated by preparative, enzymatic, and physicochemical methods. They form an intrinsic component of the histological structure and of the physiological activity of muscle tissue and are thus a common denominator of the chemical, structural, and functional aspects of a developing tissue. In this section of the present discussion, an attempt is made to review data which define quantitatively certain aspects of the development of the leg musculature of the chick embryo, and to relate the data for net protein increase to amino acid incorporation into specific muscle proteins and to the changes in some constituents of the protein-forming apparatus of the developing muscle tissue itself.

In considering protein accumulation in developing muscle, one can turn first to the relation of the increase in protein to the increase in other tissue components. As the leg musculature increases in total mass, the relation between fresh weight, extracellular space, cellular space, cell number, and dry weight changes very markedly (Herrmann *et al.*, 1957). These changes are graphically represented for the 12th to 24th day of development in Fig. 6. During this period the wet weight increases linearly. At the same time, the proportion of extracellular space in the total wet weight decreases markedly and the proportion of intracellular space increases correspondingly during this period of development. Such an increase in cell space can be due to an increase in cell number or an increase in cell size. The data given in Fig. 6 show that the cell number given as DNA increases only slightly in proportion to the wet weight during the 12–18-day period and actually declines during the 18–24-day period. This means that the cell number increases only slightly faster than the fresh weight during the 12–18-day period. During the 18–24-day period, the decline in the proportion of DNA (cell number) is inversely

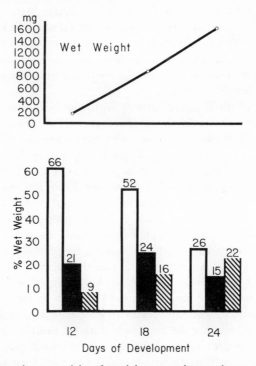

Fig. 6. *Upper graph:* wet weight of total leg musculature; *lower graph:* extracellular space (rimmed rectangles), DNA ($\times 10^2$) (solid rectangles), and dry weight (diagonally striped rectangles) in per cent of fresh weight during the development of the chick embryo from the 12th to the 24th day of development. Cellular space can be calculated as 100 − extracellular space.

proportional to the increase in fresh weight, which indicates that no increase in the absolute amount of DNA (cell number) takes place during this time and cell proliferation comes to a standstill. Thus, the changes in cell number cannot account for the increase in the proportion in cell space. However, the proportion of dry weight to wet weight increases both during the 12–18-day and the 18–24-day period, and this increase in dry weight is inversely proportional to the decrease in cell space. It has to be assumed, therefore, that the increase in cell space in proportion to the total tissue volume (wet weight) must be due to an increase in cell size and an increase in cell dry weight. The straight line increase in fresh weight shown in Fig. 6 is thus based on two different cell mechanisms. During the 12–18-day period, the fresh weight increase consists in an increase in cell number and in cell size, but it is the increase in cell size which contributes the main portion of the increase in cell space and in the increased proportion of dry weight. After the 18th day of development there is no further increase in cell number, at least during the period from the 18th to the 24th day of development. A continued replacement of extracellular space by cellular space and the continuing increase in dry weight in proportion to fresh weight are due during this period solely to an increase in cell size.

The cessation of proliferative activity with continued cell growth on the 18th day indicates a change in the mechanism of growth similar to the transition in corneal growth. In accordance with the findings for corneal development, a change in the proportions of specific cell proteins and total protein is found in muscle during this transition period from the 12th day of embryonic development to some time after hatching. From the graph in Fig. 7 it can be seen that an increasing proportion of the total protein moiety is made up of proteins of the actomyosin group. A similar, proportionately more rapid increase has been found in developing muscle tissue for the accumulation of collagen (Herrmann and Barry, 1955), but not for phosphatase (Konigsberg and Herrmann, 1955). An extrapolation of the relative myosin quantities indicated by the bar graphs in Fig. 7 shows for both methods of myosin determination a bisecting of the abscissa on the 11th day. At this date the myosin protein accumulation seemingly begins to exceed total protein production.

As a first step in relating the analytical data for protein formation to the actual activity of the protein-forming system of the muscle cell, the accumulation of the actomyosin fraction was compared at the 14th and 24th days of development with the rate of incorporation of glycine-1-C^{14} into the actomyosin fraction (Herrmann *et al.*, 1958) at the same period of development. As reference basis we adopted the expression of the determined magnitudes per DNA. In most tissues this indicates the activity levels per cell, but in muscle

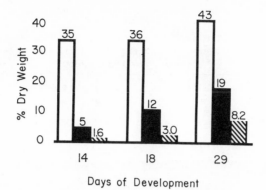

Fig. 7. Protein content (after hot trichloroacetic acid extraction) (rimmed rectangles) actomyosin + myosin content determined viscosimetrically (Visc.) (solid rectangles), and myosin content obtained by cellulose column chromatography (Prep.) (diagonally striped rectangles) in per cent of dry weight in the leg musculature of the chick embryo from the 14th day to the 29th day of development.

tissue no cells exist in the strict sense and the term nucleocytoplasmic unit (NCU) is substituted.* Calculations for the actomyosin production per NCU (DNA) showed a value of 1.37 for the 16th day and 1.65 for the 24th day of development. From Table III it can be seen that calculations per NCU (DNA) of incorporation of glycine-1-C¹⁴ into the glycine fraction of actomyosin show also very similar values for these 2 days of development. This indicates a uniform rate of actomyosin production per cell both during a period of development when proliferation is still in rapid progress, and when proliferation has markedly declined. It should be mentioned that in the tracer experiments the concentration of tracer in the free cell pool was kept at about the same level in the 14-day embryo and the 24-day chick. For this reason, the incorporation values for the two periods of development are directly comparable.

As the next step in the analysis of the protein-forming system of developing muscle, data have been obtained by Dr. Martha Tootle (Tootle, 1963) on amino acid activation and on the composition of subcellular fractions prepared from muscle during the described developmental period of the

* An expression per cell or per NCU refers directly to the activity of the protein-forming system of the cell. This avoids misleading calculations based on weight units of total or specific proteins which occur at widely varying concentrations in developing cells. If a cell produces a constant amount of a protein per day, giving a quantity of 1 N on the first and a quantity of 5 N on the fifth day, calculation of rate of production based on amount of protein present would give on the first day 5 times higher values than on the fifth day. The values calculated per DNA would remain constant.

TABLE III

INCORPORATION IN ACTOMYSIN FRACTION OF GLYCINE-1-C^{14} CALCULATED PER
ACTOMYOSIN GLYCINE (I) AND PER DNA (III) AT THE 14TH AND
24TH DAYS OF DEVELOPMENT

	Days of development	
	14	24
I Incorporation calculated as glycine-1-C^{14}/glycine-C^{12} of actomyosin fraction	27	4
II Actomyosin per DNA	3.1	18
III Incorporation per DNA (I × II)	84	72

chick embryo. The data listed in Table IV for amino acid activation, as measured colorimetrically by hydroxamate formation, show that, if calculated per DNA content, the total amino acid activating capacity of muscle tissue does not change significantly during the period under consideration. In the developing muscle, a constant value for amino acid activation per NCU would seem compatible with protein accumulation since the increase in protein per NCU per day is also nearly constant.

On fractionation of the muscle homogenates, it was found (Table V) that more than 95% of the amino acid activation activity of the homogenate

TABLE IV

AMINO ACID ACTIVATION IN HOMOGENATES OF CHICK LEG MUSCLE AT
DIFFERENT STAGES OF DEVELOPMENT[a,b]

Days of development	Millimoles hydroxamate/ hr/mg DNA × 10^{-3})
11-day embryonic	2.03
14-day embryonic	1.59
4-day hatched	2.12
8-day hatched	2.42

[a] Adapted from M. L. Tootle (1963).
[b] Heavy particle fraction removed by centrifugation at 20,000 g for 20 minutes. Amino acid activation is measured as hydroxamate formation per DNA (per cell).

TABLE V

Amino Acid Activation in Subfractions of Chick Leg Muscle at Different Stages of Development[a,b]

Days of development	Micromoles hydroxamate/mg protein N/hr)				
	S_2[c]	SCF I[d]	SCF II[e]	pH 5[f]	S_4[g]
11-day embryonic	0.17	0.067	0.41	0.09	0.09
14-day embryonic	0.21	0.025	0.55	0.17	0.23
4-day hatched	0.13	0.011	0.47	0.16	0.085
8-day hatched	0.12	0.0085	0.32	0.15	0.086

[a] Adapted from M. L. Tootle (1963).

[b] The figures indicate specific activity of hydroxamate formation per mg protein nitrogen.

[c] Supernatant obtained after removal of heavy particles by centrifugation at 20,000 g for 20 minutes and removal of SCF I fraction.

[d] Fraction obtained by centrifugation at 105,000 g for 2 hours after removal of heavy particles.

[e] Fraction obtained by centrifugation at 105,000 g for 4 hours after removal of SCF I fraction.

[f] Precipitate obtained at pH 5 from supernatant after removal of SCF II.

[g] Supernatant obtained after removal of pH 5 fraction.

could be recovered in the supernatant obtained after 2-hour centrifugation a 105,000g. About one-half of this activity could be precipitated at pH 5 as in other conventional preparations of the amino acid activating enzymes. Some activity was found also in the sediment of a subcellular fraction (SCF I) obtained on centrifugation for 2 hours at 105,000g after preceding removal of larger particles, and greater activity was obtained in a sediment from the supernatant obtained when centrifuged for an additional 4 hours at the same speed. The first sediment, SCF I (105,000g, 2 hours; sarcotubular fraction, Muscatello *et al.*, 1961), has the enzymatic properties of microsomes such as phosphatase activities and cytochrome c reductase activity, while the second fraction shows no indication of these enzymatic reactions. Therefore, the second fraction appears to be qualitatively different and is designated as a separate subcellular fraction (SCF II). The highest specific activity for amino acid activation was found in SCF II. During the 12–25-day period of development the specific activity of this fraction remained within the same order of magnitude, and only in the fraction obtained from adult animals was there a marked decrease in specific activity. The SCF I showed also some amino acid activation, but the specific activity was very low and can be attributed probably to contamination by the SCF II and/or adsorption of the activating enzymes.

Further analysis of the SCF I and SCF II fractions showed changes in

TABLE VI[a]

PROTEIN NITROGEN (A),[b] RNA (B),[b] AND PHOSPHOLIPID PHOSPHORUS[c] OF
SUBCELLULAR FRACTIONS OF CHICK LEG MUSCLE AT THREE STAGES OF
DEVELOPMENT

	SCF I[d]			SCF I$_L$[e]			SCF I$_H$[f]			SCF II[g]		
	A	B	C	A	B	C	A	B	C	A	B	C
11-day embryonic	1.26	0.53	0.20	0.208	0.067	7.2	0.726	0.452	8.7	0.177	0.022	0.7
14-day embryonic	1.32	0.63	0.26	0.473	0.154	17.9	0.548	0.427	6.8	0.213	0.024	0.5
4-hatched	0.908	0.096	0.11	0.402	0.037	8.3	0.415	0.068	3.6	0.168	0.013	0.3

[a] Adapted from M. L. Tootle (1963).
[b] Given as μg/DNA.
[c] Given as picagrams/DNA.
[d] Fraction obtained by centrifugation at 105,000 g for 2 hours after removal of heavy particles.
[e] Light SCF I subfraction: pellicle obtained at 105,000 g, 20 hours, 56% sucrose.
[f] Heavy SCF I subfraction: sediment obtained at 105,000 g, 20 hours, 56% sucrose.
[g] Fraction obtained by centrifugation at 105,000 g for 4 hours after removal of SCF I fraction.

composition during the investigated period of muscle development (Table VI). The SCF I was further fractionated into a lighter SCF I_L and heavier SCF I_H subfraction by centrifugation in 56% sucrose solution for 20 hours at 105,000g, following the procedure of Chauveau *et al.* (1962). Under these conditions SCF I_L appears as a pellicle on the surface of the suspending solution, while SCF I_H, much richer in RNA, settles as a pellet to the bottom of the centrifuge tube. Both the yield per DNA and the changes in composition in these fractions can be seen in Table VI. The total protein content of the SCF I does not change during the 11–14-day period and declines only slightly up to the 24th day. The RNA shows a distinct decline during the 11–14-day period and a much steeper drop during the later period. The phospholipid phosphorus content of the SCF I remains practically unchanged during the first period and shows a distinct drop during the later period of development. The data for the SCF I mask in part divergent trends in the heavy and light subfractions. In SCF I_L there is a marked increase in all components in the first period of development. In the second part, the protein content declines only slightly, the RNA shows again a sharp drop, and also the phospholipid phosphorus fraction declines markedly. The protein and phospholipid phosphorus contents of SCF I_H show a steady decline, RNA stays on about the same level during the first part of the developmental period and shows again the same sharp drop as the other fractions during the late part of development.

The changes in the SCF II are not very characteristic. The elevation in protein nitrogen on the 14th day is hardly significant, and the decline in RNA is less extensive than in the other fractions. The phospholipid phosphorus content declines in the same proportion as in the SCF I.

C. *Discussion*

In the second part of this paper an attempt was made to define in quantitative terms the transition of certain embryonic tissues from a phase of rapid cell proliferation, with production of specific cell constituents at a slow rate, to a phase of declining proliferation and increased production of specific cell proteins. In view of the recent data obtained by Stockdale and Holtzer (1961), it seems likely that during this transition period synthesis of DNA and of proteins needed in the process of cell proliferation comes to an end in an increasing number of cells, but synthesis of proteins which are specific cell products continues at about the same rate. The abruptness of this transition is indicated, at least for muscle cells *in vitro*, by the experiments of Stockdale and Holtzer (1961). A similar effect can be observed in myoblast cultures on inhibition of DNA synthesis by nitrogen mustard (Konigsberg *et al.*, 1960). Both in muscle and in the cornea, at least a temporary cessation of DNA formation occurs between the 16th and 18th day of development. The possibility that the decline in DNA accumulation in muscle is due, at least in part, to hormonal control is suggested by the data of Konigsberg (1958) and Love and Konigsberg (1958).

The rapid accumulation of myosin fractions in muscle has been demonstrated with different methods for myosin determination. It should be pointed out that the preparative method indicated in Fig. 7 yields myosin of high purity and the properties of this fraction did not seem to change during development. More detailed evidence for the absence of significant differences of embryonic and adult myosin will be published elsewhere (Love, 1960).

The transition from a phase of synthesis of nonspecific to one of synthesis of specific proteins seems to involve a shift in the quality of proteins synthesized rather than a change in the over-all rate of protein formation. Therefore, it seemed of interest to find out whether any change could be detected during this period by a direct analysis of the protein-forming centers of the muscle cell. The most striking changes observed so far are the changes in the composition of the subcellular fractions, in particular of the light and heavy subfractions of SCF I, which may form components of the sarcoplasmic reticulum. The assumption that these light and heavy subfractions represent the membrane and ribosomal components, respectively, is borne out by the relatively high RNA content of the heavy fraction and the higher phospho-lipid-phosphate content of the light fraction during the later part of develop-ment, as well as by the sedimentation behavior of the fractions. If this assump-tion is accepted, the data for the light fraction would give a quantitative index for the enlargement of the membrane portion of the endoplasmic reticulum which has been observed electron microscopically in maturing tissues (Schulze, 1961; Hay, 1958; Slautterback and Fawcett, 1959). There-fore, it is of interest to find in muscle tissue an increase in the membrane component just at the time when production of actomyosin begins to exceed significantly the production of the other proteins.

Another striking change in the composition of the SCF I, SCF I_L, and SCF I_H is the sharp decline in RNA content during the later phase of develop-ment. However, the interpretation of this change is uncertain at this time. The possibility has to be considered that a subfraction with similar sedimenta-tion properties but with smaller RNA content increases at this time of devel-opment. A heterogeneity of this subfraction could arise, perhaps in connection with the increasing structural complexity of the developing muscle cell. A further separation of the components of this fraction and a clarification of this uncertainty is being attempted. Further analysis of the changes in the protein-forming system of the embryonic chick muscle during the investigated period of development seems desirable in view of other data. Konigsberg (1958) suggested that protein synthesis after the 16th day of muscle develop-ment is maintained by a more rapid turnover of a smaller quantity of RNA. Before the 16th day, a larger and more stable RNA fraction would lead to synthesis of about equal amounts of protein. This suggestion was made on the basis of Mazia and Prescott's (1955) assumption of a dual mechanism of RNA function in protein synthesis. This suggestion should be reconsidered in the light of the present, more detailed information about RNA fractions involved in protein synthesis. Of interest are recent observations of changes in the composition of ribosomes at different growth phases of cells (Loening, 1961; Rabson *et al.*, 1961), and in the relative rates of incorporation of amino acid into proteins of different RNA fractions of the nucleus during the transi-tion from rapid proliferation to a nuclear "resting" state (Harbers, 1961). Changes in microsomal composition have been noted during the different phases of liver regeneration (Hultin and von der Decken, 1958; McCorquo-dale *et al.*, 1961). Other observations of changes in the protein-forming system during development have been referred to in the introduction to this paper.

From the point of view of developmental biology, it is of interest that many data point to the 16–18-day period in chick development as a phase of marked biochemical changes. The decline in DNA accumulation in muscle and in the connective tissues of the eye was noted before. The replacement of heart lactic dehydrogenase by skeletal muscle lactic dehydrogenase in muscle tissue is accelerated during this period (Cahn *et al.*, 1962). Obviously the findings discussed in the second section of this paper deal only with one particular step in the readjustment of cellular mechanisms during development. The muscle, just as most other embryonic cells, passes undoubtedly through several critical phases. It should be recalled that in many species muscle development becomes dependent at some time of development upon interaction with the nerve axons, and failure of innervation leads to muscle cell degeneration. This apparent instability of embryonic cells can be regarded as a deplorable liability because biochemical findings pertaining to one phase of development may not apply to the same tissue at the following day of embryogenesis. However, this liability can be regarded as a virtue. It is this changing pattern in molecular mechanism during development which affords a unique opportunity for comparative studies, for the investigation of controlling mechanisms of genic, hormonal, or other physiological nature, and for the understanding of the relation of molecular, morphological, and functional properties which make up the totality of events which we call loosely a mature cell.

It is in this sense that it may be justified to paraphrase the introductory motto as follows:

Each stirring turn of time has a second's rest between motions,
Which reveals the relation of parts, and past and future as one.

Acknowledgments

The author is Maud K. Irving Research Professor of the American Cancer Society. Original work reported in this paper was carried out with the aid of grants E-549 and B-2238 of the National Institute of Neurological Diseases and Blindness of the United States Public Health Service, and of grants from the Muscular Dystrophy Association and the Association for Aid of Crippled Children. This article is contribution No. 73 of the Institute of Cellular Biology, University of Connecticut, Storrs, Connecticut. The author is indebted to Dr. M. L. Tootle and Dr. J. R. Coleman for suggestions and help in the preparation of the manuscript.

References

ACKERMANN, W. W., AND TAYLOR, A. (1948). Application of a metabolic inhibitor to the developing chick embryo. *Proc. Soc. Exptl. Biol. Med.* **67,** 449–452.

ALLFREY, V. G., HOPKINS, J. W., FRENSTER, J. H., AND MIRSKY, A. E. (1960). Reactions governing incorporation of amino acids into the proteins of the isolated cell nucleus. *Ann. N.Y. Acad. Sci.* **88,** 722–740.

BREITMAN, T. R., AND WEBSTER, G. C. (1958). Effect of chloramphenicol on protein and nucleic acid synthesis in isolated thymus nuclei. *Biochim. et Biophys. Acta* **27**, 408–409.

BRITT, G. L., AND HERRMANN, H. (1959). Protein accumulation in early chick embryos grown under different conditions of explanation. *J. Embryol. Exptl. Morphol.* **7**, 66–72.

BUCKINGHAM, B. J., AND HERRMANN, H. In preparation.

CAHN, R. D., KAPLAN, N. O., LEVINE, L., AND ZWILLING, E. (1962). Nature and development of lactic dehydrogenases. *Science* **136**, 962–969.

CHAUVEAU, J., MOULÉ, Y., ROULLIER, C., AND SCHNEEBELI, J. (1962). Isolation of smooth vesicles and free ribosomes from rat liver microsomes. *J. Cell Biol.* **12**, 17–29.

CHRISTENSEN, H. N. (1959). Active transport with special reference to the amino acids. *Perspectives in Biol. Med.* **2**, 228–242.

CHRISTENSEN, H. N., STREICHER, J. S., ROTHWELL, J., AND SEARS, R. A. (1948). Association between rapid growth and elevated cell concentrations of amino acids. *J. Biol. Chem.* **175**, 96–105.

COHEN, G. N., AND RICKENBERG, H. V. (1956). Concentration spécifique réversible des amino acides chez *Escherichia coli*. *Ann. inst. Pasteur* **91**, 693–720.

COHEN, G. N., HALVORSON, H. O., AND SPIEGELMAN, S. (1958). Effects of *p*-fluorophenyl-alanine on the growth and physiology of yeast. *In* "Microsomal Particles and Protein Synthesis" (R. B. Roberts, ed.), pp. 100–108. Pergamon Press, New York.

COHN, M. (1957). Contributions of studies on the β-galactosidase of *Escherichia coli* to our understanding of enzyme synthesis. *Bacteriol. Revs.* **21**, 140–168.

COWIE, D. B., AND MCCLURE, F. T. (1959). Metabolic pools and the synthesis of macromolecular molecules. *Biochim. et Biophys. Acta* **31**, 236–245.

COWIE, D. B., COHEN, G. N., BOLTON, E. T., AND DE ROBICHON-SZULMAJSTER, H. (1959). Amino acid analog incorporation into bacterial proteins. *Biochim. et Biophys. Acta* **34**, 39–46.

CSAPO, A., AND HERRMANN, H. (1951). Quantitative changes in contractile proteins of chick skeletal muscle during and after embryonic development. *Am. J. Physiol.* **165**, 701–710.

DAGLEY, S., AND SYKES, J. (1960). Bacterial ribonucleoprotein synthesized in the presence of chloramphenicol. *Biochem. J.* **74**, 11 P.

DEUCHAR, E. M. (1960). The distribution of free leucine and its uptake into protein in *Xenopus laevis* embryos. *Symposium on Germ Cells and Development (Inst. intern. Embryol. and Fondazione A. Baselli)* pp. 537–544.

DEUCHAR, E. M. (1961). Amino acid activation in embryonic tissues of *Xenopus laevis*. I. Increased ^{32}P exchange between pyrophosphate and adenosine triphosphate in the presence of added L-leucine. *Exptl. Cell Research* **25**, 364–373.

DEUCHAR, E. M. (1962). Amino acid activation in embryonic tissues of *Xenopus laevis*. II. Hydroxamic acid formation in the presence of L-leucine. *Exptl. Cell Research* **26**, 568–570.

DEUCHAR, E. M., AND HERRMANN, H. (1962). Uptake of amino acids into explanted chick embryos by epidermal and endodermal routes. *Acta Embryol. Morphol. Exptl.* **5**, 161–166.

EAGLE, H., PIEZ, K. A., FLEISCHMAN, R., AND OYAMA, V. I. (1959). Protein turnover in mammalian cell cultures. *J. Biol. Chem.* **234**, 592–597.

EBERT, J. D. (1953). An analysis of the synthesis and distribution of the contractile protein myosin in the development of the heart. *Proc. Natl. Acad. Sci. U.S.* **39**, 333–344.

EBERT, J. D. (1955). Some aspects of protein biosynthesis in development. *In* "Aspects of Synthesis and Order in Growth," Growth Symposium No. 13 (D. Rudnick, ed.), pp. 69–112. Princeton Univ. Press, Princeton, New Jersey.

FELDMAN, M., AND WADDINGTON, C. H. (1955). The uptake of methionine-S[35] by the chick embryo and its inhibition by ethionine. *J. Embryol. Exptl. Morphol.* **3,** 44–58.

FLEXNER, L. B., FLEXNER, J. B., AND ROBERTS, R. B. (1958). Biochemical and physiological differentiation during morphogenesis. *J. Cellular Comp. Physiol.* **51,** 385–403.

FRAENKEL, D. G., AND NEIDHARDT, F. C. (1961). Use of chloramphenicol to study control of RNA synthesis in bacteria. *Biochim. et Biophys. Acta* **53,** 96–110.

FRISCH, L., ed. (1961). Cellular regulatory mechanisms. *Cold Spring Harbor Symposia Quant. Biol.* **26.**

GALE, E. F. (1958). The mode of action of chloramphenicol. *Ciba Foundation Symposium on Amino Acids and Peptides with Antimetabolic Activity,* pp. 19–34.

GIUDICE, G., AND MONROY, A. (1958). Incorporation of S[35]-methionine in the proteins of the mitochondria of developing and parthenogenetically activated sea urchin eggs. *Acta Embryol. Morphol. Exptl.* **2,** 58–65.

GIUDICE, G., VITTORELLI, M. L., AND MONROY, A. (1962). Investigations on the protein metabolism during the early development of the sea urchin. *Acta Embryol. Morphol. Exptl.* **5,** 113–122.

GOLDBERG, I. H., AND RABINOWITZ, M. (1962). Actinomycin D inhibition of deoxyribonucleic acid-dependent synthesis of ribonucleic acid. *Science* **136,** 315–316.

GORSKI, J., AIZAWA, Y., AND MUELLER, G. C. (1961). Effect of puromycin *in vivo* on the synthesis of protein, RNA and phospholipids in rat tissues. *Arch. Biochem. Biophys.* **95,** 508–511.

HALVORSON, H. (1958a). Intracellular protein and nucleic acid turnover in resting yeast cells. *Biochim. et Biophys. Acta* **27,** 255–266.

HALVORSON, H. (1958b). Studies on protein and nucleic acid turnover in growing cultures of yeast. *Biochim. et Biophys. Acta* **27,** 267–276.

HANCOCK, R. (1960). Accumulation of pool amino acids in *Staphylococcus aureus* following inhibition of protein synthesis. *Biochim. et Biophys. Acta* **37,** 47–55.

HARBERS, E. (1961). Untersuchungen der Ribonukleinsäure an Zelkernen. II. Stoffwechselverhalten verschiedener Ribonukleinsäure-Fraktionen aus den Kernen "ruhender" und proliferierender Zellen. *Z. physiol. Chem., Hoppe-Seyler's* **327,** 3–12.

HAY, E. (1958). The fine structure of blastema cells and differentiating cartilage cells in regenerating limbs of Amblystoma larvae. *J. Biophys. Biochem. Cytol.* **4,** 583–592.

HAYASHI, Y., AND HERRMANN, H. (1959). Growth and glycine incorporation in chick embryo explants. *Develop. Biol.* **1,** 437–458.

HERRMANN, H. (1952). Studies of muscle development. *Ann. N. Y. Acad Sci.* **55,** 99–108.

HERRMANN, H. (1953). Interference of amino acid analogues with normal embryonic development. *J. Embryol. Exptl. Morphol.* **1,** 291–295.

HERRMANN, H. (1957). Protein synthesis and tissue integrity in the cornea of the developing chick embryo. *Proc. Natl. Acad. Sci. U.S.* **43,** 1007–1011.

HERRMANN, H. (1958). Some problems of protein formation in the sclera and cornea of the chick embryo. *In* "The Chemical Basis of Development" (W. D. McElroy and B. Glass, eds.), pp. 329–338. Johns Hopkins Press, Baltimore, Maryland.

HERRMANN, H. (1960). Molecular mechanism of differentiation. *In* "Fundamental Aspects of Normal and Malignant Growth" (W. W. Nowinski, ed.), pp. 495–545. Elsevier, Amsterdam.

HERRMANN, H. (1961). Tissue interaction and differentiation in the corneal and scleral stroma. *In* "The Structure of the Eye" (G. Smelser, ed.), pp. 421–433. Academic Press, New York.

HERRMANN, H., AND BARRY, S. R. (1955). Accumulation of collagen in skeletal muscle, heart and liver of the chick embryo. *Arch. Biochem. Biophys.* **55**, 526–533.

HERRMANN, H., AND LEBEAU, P. L. (1962). ATP level, cell injury, and apparent epithelium-stroma interaction in the cornea. *J. Cell Biol.* **13**, 465–467.

HERRMANN, H., AND MARCHOK, A. (1963). Gain and loss of protein in explanted chick embryos. *Develop. Biol.* **7**, 207–218.

HERRMANN, H., AND NICHOLAS, J. S. (1948a). Quantitative changes in muscle protein fractions during rat development. *J. Exptl. Zool.* **107**, 165–176.

HERRMANN, H., AND NICHOLAS, J. S. (1948b). Enzymatic liberation of inorganic phosphate from adenosine triphosphate in developing rat muscle. *J. Exptl. Zool.* **107**, 177–182.

HERRMANN, H. AND SCHULTZ, P. W. (1958). Incorporation of glycine into the proteins of explanted chick embryos. *Arch. Biochem. Biophys.* **73**, 296–305.

HERRMANN, H., AND TOOTLE, M. L. *Physiol. Revs.* In preparation.

HERRMANN, H., KONIGSBERG, U. R., AND CURRY, M. F. (1955). A comparison of the effects of antagonists of leucine and methionine on the chick embryo. *J. Exptl. Zool.* **128**, 359–378.

HERRMANN, H., WHITE, B. N., AND COOPER, M. (1957). The accumulation of tissue components in the leg muscle of the developing chick. *J. Cellular Comp. Physiol.* **49**, 227–251

HERRMANN, H., LERMAN, L., AND WHITE, B. N. (1958). Uptake of glycine-1-^{14}C into the actomyosin and collagen fractions of developing chick muscle. *Biochim. et Biophys. Acta* **27**, 161–164.

HOAGLAND, M. B. (1960). The relationship of nucleic acid and protein synthesis as revealed by studies in cell-free systems. *In* "The Nucleic Acids" (E. Chargaff and J. N. Davidson, eds.), Vol. 3, pp. 349–408. Academic Press, New York.

HOLTZER, H. (1961). Aspects of chondrogenesis and myogenesis. *In* "Synthesis of Molecular and Cellular Structure," Growth Symposium No. 19 (D. Rudnick, ed.), pp. 35–87. Ronald Press, New York.

HOLTZER, H., MARSHALL, J. M., JR., AND FINCK, H. (1957). An analysis of myogenesis by the use of fluorescent antimyosin. *J. Biophys. Biochem. Cytol.* **3**, 705–724.

HULTIN, T. (1952). Incorporation of N^{15} labeled glycine and alanine into the proteins of developing sea urchin eggs. *Exptl. Cell Research* **3**, 494–496.

HULTIN, T. (1953a). Incorporation of N^{15}-DL-alanine into protein fractions of sea urchin embryos. *Arkiv Kemi* **5**, 559–564.

HULTIN, T. (1953b). Metabolism and determination. *Arch. néerl. zool.* **10** (Suppl.), 76–91.

HULTIN, T. (1961). Activation of ribosomes in sea urchin eggs in response to fertilization. *Exptl. Cell Research* **25**, 405–417.

HULTIN, T., AND BERGSTRAND, A. (1960). Incorporation of C^{14}-L-leucine into protein by cell-free systems from sea urchin embryos at different stages of development. *Develop. Biol.* **2**, 61–75.

HULTIN, T., AND VON DER DECKEN, A. (1958). The activity of soluble cytoplasmic constituents from regenerating rat liver in amino acid incorporating systems. *Exptl. Cell Research* **15**, 581–594.

INGRAM, V. M. (1962). The genetic control of protein specificity. *In* "The Molecular Control of Cellular Activity" (J. M. Allen, ed.), pp. 179–188. McGraw-Hill, New York.

JACOB, F., AND MONOD, J. (1961). Genetic regulatory mechanisms in the synthesis of proteins. *J. Mol. Biol.* **3**, 318–356.

KEMPNER, E. S., AND COWIE, D. B. (1960). Metabolic pools and the utilization of amino acid analogs for protein synthesis. *Biochim. et Biophys. Acta* **42**, 401–408.

KERSTEN, W., AND KERSTEN, H. (1962). Zur Wirkungsweise von Actinomycinen. II. Bildung überschüssiger Desoxyribonucleinsäure in Bacillus subtilis. *Z. physiol. Chem. Hoppe-Seyler's* **327**, 234–242.

KLEIN, N., McCONNELL, E., AND BUCKINGHAM, B. J. (1962a). Growth of explanted chick embryos on a chemically defined medium and effects of specific amino acid deficiencies. *Develop. Biol.* **5**, 296–308.

KLEIN, N., McCONNELL, E., AND RIQUIER, D. J. (1962b). Effects of high oxygen levels on the growth and survival of explanted chick embryos. *Am. Zool.* **2**, 420.

KONIGSBERG, I. R. (1958). Thyroid regulation of protein and nucleic acid accumulation in developing skeletal muscle of the chick embryo. *J. Cellular Comp. Physiol.* **52**, 13–41.

KONIGSBERG, I. R., AND HERRMANN, H. (1955). The accumulation of alkaline phosphatase in developing chick muscle. *Arch. Biochem. Biophys.* **55**, 534–545.

KONIGSBERG, I. R., McELVAIN, N., TOOTLE, M., AND HERRMANN, H. (1960). The dissociability of deoxyribonucleic acid synthesis from the development of multinuclearity of muscle cells in culture. *J. Biophys. Biochem. Cytol.* **8**, 333–343.

LANDAUER, W. (1957). Niacin antagonists and chick development. *J. Exptl. Zool.* **136**, 509–530.

LANDAUER, W. (1960). Nicotine-induced malformations of chicken embryos and their bearing on the phenocopy problem. *J. Exptl. Zool.* **143**, 107–122.

LEBLOND, C. P., AND AMANO, M. (1962). Symposium: Synthetic processes in the cell nucleus. IV. Synthetic activity in the nucleolus as compared to that in the rest of the cell. *J. Histochem. and Cytochem.* **10**, 162–174.

LIPKIN, M., ALMY, T. P., AND QUASTLER, H. (1961). Stability of protein in intestinal epithelial cells. *Science* **133**, 1019–1021.

LOENING, U. E. (1961). Changes in microsomal components accompanying cell differentiation of pea-seedling roots. *Biochem. J.* **81**, 254–260.

LOVE, D. S. (1960). Preparation and purification of embryonic chick myosin. *Federation Proc.* **19**, 256.

LOVE, D. S., AND KONIGSBERG, I. R. (1958). Enhanced desoxyribonucleic acid accumulation and retarded protein accumulation in skeletal muscle of "hypophysectomized" chick embryos. *Endocrinology* **62**, 378–384.

McCORQUODALE, D. J., VEACH, E. G., AND MUELLER, G. C. (1961). The incorporation *in vitro* of labeled amino acids into the proteins of normal and regenerating rat liver. *Biochim. et Biophys. Acta* **46**, 335–343.

McKENZIE, J., AND EBERT, J. D. (1960). The inhibitory action of antimycin A in the early chick embryo. *J. Embryol. Exptl. Morphol.* **8**, 314–320.

McMASTER-KAYE, R. (1962). Symposium: Synthetic processes in the cell nucleus. III. The metabolism of nuclear ribonucleic acid in salivary glands of *Drosophila repleta. J. Histochem. and Cytochem.* **10**, 154–161.

MANDELSTAM, J., AND HALVORSON, H. (1960). Turnover of protein and nucleic acid in soluble and ribosome fractions of non-growing *Escherichia coli. Biochim. et Biophys. Acta* **40**, 43–49.

MATTHAEI, J. H., AND NIRENBERG, M. W. (1961). Characteristics and stabilization of DNAase-sensitive protein synthesis in *E. coli* extracts. *Proc. Natl. Acad. Sci. U.S.* **47**, 1580–1588.

MAZIA, D., AND PRESCOTT, D. M. (1955). The role of the nucleus in protein synthesis in amoeba. *Biochim. et Biophys. Acta* **17**, 23–34.

MEYER, K., AND CHAFFEE, E. (1940). The mucopolysaccharide acid of the cornea and its enzymatic hydrolysis. *Am. J. Ophthalmol.* **23**, 1320–1324.

MONROY, A. (1960). Incorporation of S^{35}-methionine in the microsomes and soluble proteins during the early development of the sea urchin egg. *Experientia* **16**, 114–115.

MONROY, A., VITTORELLI, M. L., AND GUARNERI, R. (1961). Investigations on the proteins of the cell fluid during the early development of the sea urchin *Paracentrotus lividus*. *Acta Embryol. Morphol. Exptl.* **4**, 77–95.

MORRIS, A., FAVELUKES, S., ARLINGHAUS, R., AND SCHWEET, R. (1962). Mechanism of puromycin inhibition of hemoglobin synthesis. *Biochem. Biophys. Research Communs.* **7**, 326–330.

MUNIER, R., AND COHEN, G. N. (1959). Incorporation d'analogues structuraux d'amino acides dans les protèines bactériennes au cours de leur synthèse *in vivo*. *Biochim. et Biophys. Acta* **31**, 378–391.

MUSCATELLO, U., ANDERSSON-CEDERGREN, E., AZZONE, G. F., AND VON DER DECKEN, A. (1961). The sarcotubular system of frog skeletal muscle. A morphological and biochemical study. *J. Biophys. Biochem. Cytol.* **10**, 201–218.

NAKANO, E., AND MONROY, A. (1958). Incorporation of S^{35}-methionine in the cell fractions of sea urchin eggs and embryos. *Exptl. Cell Research* **14**, 236–244.

NASS, M. M. K. (1962). Developmental changes in frog actomyosin characteristics. *Develop. Biol.* **4**, 289–320.

NATHANS, D., VON EHRENSTEIN, G., MONRO, R., AND LIPMANN, F. (1962). Protein synthesis from aminoacyl-soluble ribonucleic acid. *Federation Proc.* **21**, 127–133.

NEEDHAM, J. (1942). "Biochemistry and Morphogenesis," p. 505. Cambridge Univ. Press, London and New York.

NEWBURGH, R. W., BUCKINGHAM, B. J., AND HERRMANN, H. (1962). Levels of reduced TPN generating systems in chick embryos *in ovo* and in explants. *Arch. Biochem. Biophys.* **97**, 94–99.

NIRENBERG, M. W., AND MATTHAEI, J. H. (1961). The dependence of cell-free protein synthesis in *E. coli* upon naturally occurring or synthetic polyribonucleotides. *Proc. Natl. Acad. Sci. U.S.* **47**, 1588–1602.

O'BRIEN, B. R. A. (1961). Development of haemoglobin by de-embryonated chick blastoderms cultured *in vitro* and the effect of abnormal RNA upon its synthesis. *J. Embryol. Exptl. Morphol.* **9**, 202–221.

OGAWA, Y. (1962). Synthesis of skeletal muscle proteins in early embryos and re-generating tissue of chick and *Triturus*. *Exptl. Cell Research* **26**, 269–274.

PALADE, G. (1958). Microsomes and ribonucleoprotein particles. *In* "Microsomal Particles and Protein Synthesis" (R. B. Roberts, ed.), pp. 36–49. Pergamon Press, New York.

PARDEE, A. B. (1962). Aspects of genetic and metabolic control of protein synthesis. *In* "The Molecular Control of Cellular Activity" (J. M. Allen, ed.), pp. 265–278. McGraw-Hill, New York.

Piez, K. A., and Eagle, H. (1958). The free amino acid pool of cultured human cells. *J. Biol. Chem.* 231, 533–545.

Pollock, M. R. (1959). Induced formation of enzymes. In "The Enzymes" (P. D. Boyer, H. Lardy, and K. Myrbäck, eds.), 2nd ed., Vol. 1, pp. 619–680. Academic Press, New York.

Prescott, D. M. (1962). Symposium: Synthetic processes in the cell nucleus. II. Nucleic acid and protein metabolism in the macronuclei of two ciliated protozoa. *J. Histochem. and Cytochem.* 10, 145–153.

Rabinovitz, M., and Fisher, J. M. (1962). A dissociative effect of puromycin on the pathway of protein synthesis by Ehrlich ascites tumor cells. *J. Biol. Chem.* 237, 477–481.

Rabson, R., and Novelli, G. D. (1960). The incorporation of leucine-C^{14} into protein by a cell-free preparation from maize kernels. *Proc. Natl. Acad. Sci. U.S.* 46, 484–488.

Rabson, R., Mans, R. J., and Novelli, G. D. (1961). Changes in cell-free amino acid incorporating activity during maturation of maize kernels. *Arch. Biochem. Biophys.* 93, 555–562.

Rendi, R. (1959). The effect of chloramphenicol on the incorporation of labelled amino acids into proteins by isolated subcellular fractions from rat liver. *Exptl. Cell Research* 18, 187–189.

Reverberi, G., Verly, W. G., Mansueto, C., and D'Anna, T. (1960). An analysis of the ascidian development by the use of radio-isotopes. *Acta Embryol. Morphol. Exptl.* 3, 202–212.

Rickenberg, H. V., and Lester, G. (1955). The preferential synthesis of β-galactosidase in *Escherichia coli*. *J. Gen. Microbiol.* 13, 279–284.

Roberts, R. B., ed. (1958). "Microsomal Particles and Protein Synthesis." Pergamon Press, New York.

Rothfels, U. (1954). The effects of some amino acid analogues on the development of the chick embryo *in vitro*. *J. Exptl. Zool.* 125, 17–37.

Schultz, P. W. (1959). A note on Ω-bromoallyl glycine-induced pycnosis in explanted chick embryos. *Exptl. Cell Research* 17, 353–358.

Schultz, P. W., and Herrmann, H. (1958). Effect of a leucine analogue on incorporation of glycine into the proteins of explanted chick embryos. *J. Embryol. Exptl. Morphol.* 6, 262–269.

Schulze, W. (1961). Elektronenmikroskopische und histometrische Untersuchungen des Herzmuskelgewebes vom Hund während des postnatalen Wachstums. *Acta Biol. et Med. Ger.* 7, 24–31.

Sibatani, A., de Kloet, S. R., Allfrey, V. G., and Mirsky, A. E. (1962). Isolation of a nuclear RNA fraction resembling DNA in its base composition. *Proc. Natl. Acad. Sci. U.S.* 48, 471–477.

Sirlin, J. L., and Waddington, C. H. (1956). ·Cell sites of protein synthesis in the early chick embryo, as indicated by autoradiographs. *Exptl. Cell Research* 11, 197–205.

Slautterback, D. B., and Fawcett, D. W. (1959). The development of the cnidoblast of hydra. An electronmicroscope study of cell differentiation. *J. Biophys. Biochem. Cytol.* 5, 441–452.

Spiegel, M., and Frankel, D. L. (1961). Role of enzyme induction in embryonic development. *Science* 133, 275.

SPRATT, N. T., JR. (1947). Development *in vitro* of the early chick blastoderm explanted on yolk and albumen extract saline-agar substrata. *J. Exptl. Zool.* 106, 345–365.

SPRATT, N. T., JR. (1958). Chemical control of development. In "The Chemical Basis of Development" (W. D. McElroy and B. Glass, eds.), pp. 629–642. Johns Hopkins Press, Baltimore, Maryland.

STEINBERG, D., AND VAUGHAN, M. (1956a). Intracellular protein degradation *in vitro*. *Biochim. et Biophys. Acta* 19, 584–585.

STEINBERG, D., AND VAUGHAN, M. (1956b). Observations on intracellular protein catabolism studied *in vitro*. *Arch. Biochem. Biophys.* 65, 93–105.

STOCKDALE, F. E., AND HOLTZER, H. (1961). DNA synthesis and myogenesis. *Exptl. Cell Research* 24, 508–520.

TOOTLE, M. L. (1963). Investigations of amino acid activation and its relation to subcellular fractions in developing tissues. Ph.D. Thesis, University of Connecticut, Storrs, Connecticut.

VAUGHAN, M., AND STEINBERG, D. (1958). Incorporation of amino acid analogues into crystalline proteins. *Proc. Intern. Congr. Biochem., 4th, Vienna* Symposium VIII, pp. 1–15.

VAUGHAN, M., AND STEINBERG, D. (1960). Biosynthetic incorporation of fluorophenylalanine into crystalline proteins. *Biochim. et Biophys. Acta* 40, 230–236.

VOLKIN, E. (1962). Synthesis and function of the DNA like RNA. *Federation Proc.* 21, 112–119.

VON EHRENSTEIN, G., AND LIPMANN, F. (1961). Experiments on hemoglobin biosynthesis. *Proc. Natl. Acad. Sci. U.S.* 47, 941–950.

VON HAHN, H. P., AND HERRMANN, H. (1962). Effects of amino acid analogs on growth and catheptic activity of chick embryo systems. *Develop. Biol.* 5, 309–327.

WADDINGTON, C. H. (1932). Experiments on the development of chick and duck embryos cultivated *in vitro*. *Phil. Trans. Roy. Soc.* B221, 179–230.

WADDINGTON, C. H., AND SIRLIN, J. L. (1954). The incorporation of labeled amino acids into amphibian embryos. *J. Embryol. Exptl. Morphol.* 2, 340–347.

WADDINGTON, C. H., AND SIRLIN, J. L. (1959). The changing pattern of amino acid incorporation in developing mesoderm cells. *Exptl. Cell Research* 17, 582–585.

WALKER, M. S., AND WALKER, J. B. (1962). Repression of transamidinase activity during embryonic development. *J. Biol. Chem.* 237, 473–476.

WESTHEAD, E. W., AND BOYER, P. D. (1961). The incorporation of *p*-fluorophenylalanine into some rabbit enzymes and other proteins. *Biochim. et Biophys. Acta* 54, 145–156.

WILT, F. H. (1962). The ontogeny of chick embryo hemoglobin. *Proc. Natl. Acad. Sci. U.S.* 48, 1582–1590.

WINNICK, T., AND GOLDWASSER, R. (1961). Immunological investigation on the origin of myosin of skeletal muscle. *Exptl. Cell Research* 25, 428–436.

WORK, T. S. (1959). Protein biosynthesis: Some connecting links between genetics and biochemistry. In "Developing Cell Systems and Their Control," Growth Symposium No. 18 (D. Rudnick, ed.), pp. 205–235. Ronald Press, New York.

ZAMECNIK, P. C. (1960). Historical and current aspects of the problem of protein synthesis. *Harvey Lectures* 54, 256–281.

ZAMECNIK, P. C. (1962). Soluble ribonucleic acid and protein synthesis. In "The Molecular Control of Cellular Activity" (J. M. Allen, ed.), pp. 259–264. McGraw-Hill, New York.

Chromosomes and Cytodifferentiation

JOSEPH G. GALL*

Department of Zoology, University of Minnesota, Minneapolis, Minnesota

It is now widely believed that ribonucleic acid (RNA) transmits genetic information from the deoxyribonucleic acid (DNA) of the chromosomes to the sites of protein synthesis in the cytoplasm. For technical reasons much of our knowledge on the relationships between nucleic acids and proteins is derived from studies on microorganisms, although such important information as the details of hemoglobin synthesis has come from higher animals. Again for technical reasons it is generally impossible to relate specific gene functions to visible changes in the chromosomes. In most cells RNA synthesis is at a peak during the interphase stage when the chromosomes are least amenable to study by present-day techniques. Two exceptions are known, however: the giant polytene chromosomes found in larval Diptera and the even larger lampbrush chromosomes characteristic of a variety of oocytes. This paper will review recent studies on these giant chromosomes with emphasis on the morphological and cytochemical changes associated with RNA metabolism.

I. Structural Considerations

We may begin our discussion of the lampbrush and polytene chromosomes with a brief statement of their similarities and differences. More detailed information is available in several recent reviews (Gall, 1958; Beermann, 1959; Callan and Lloyd, 1960a; Callan, 1962; Swift, 1962). Both kinds of giant chromosome are considerably "unwound" relative to the condensed condition typical of a mitotic chromosome. Whether there has been intercalary growth with consequent elongation of the basic chromosome strands is not clear. But it is certain that the lampbrush chromosomes have attained their large size without replication beyond that normal for meiotic chromosomes; that is, they consist of four chromatids in two sets of two. On the other hand, the polytene chromosomes have undergone some 10 to 15 replications without separation of strands, and therefore consist of a thousand or

* The original studies reported here were supported by research grants from the National Science Foundation and the National Cancer Institute, U.S. Public Health Service.

more units comparable to a mitotic chromatid. In neither case, however, are the strands thought to be completely unwound. The chromomeres of the lampbrush chromosomes and the bands of the polytene chromosomes presumably represent areas of more tightly coiled thread, although this supposition is difficult to establish unequivocally from electron micrographs. The tightly wound condition apparently corresponds to the inactive state, as evidenced by lack of isotope incorporation or RNA accumulation in these regions. On the other hand, the threads can unravel to produce a "puff" (in the polytene chromosomes) or a "loop" (in the lampbrush chromosomes). These regions actively incorporate RNA precursors and accumulate cyto-chemically demonstrable RNA. In fact, the typical loop or puff is visible only because of its ribonucleoprotein matrix; the DNA is so dispersed in these regions that a Feulgen reaction is not usually detectable. The much discussed DNA puffs, which so far have been found in a few species only, are excep-tional in showing a disproportionate increase in DNA (Breuer and Pavan, 1955; Rudkin and Corlette, 1957; Stich and Naylor, 1958; Swift, 1962). Figures 1–4 show diagrammatically the postulated structure of the loops and puffs. From these diagrams one can see that puffing and loop formation are similar phenomena, the one occurring in a multistranded chromosome, the other in a two-stranded, prophase chromosome.

The nature of the strands which form the "backbone" of the puffs and loops is of considerable importance to our understanding of chromosome

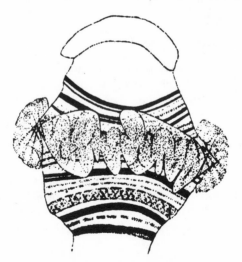

Fig. 1. A segment of *Chironomus* salivary gland chromosome showing a moderately developed puff (Balbiani ring). From Beermann (1952a).

FIG. 2. Diagram of the chromosome fibrils in the region of a puff. The chromosome would actually contain several thousand fibrils instead of the eight shown here. From Beermann (1952a).

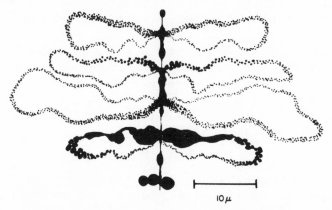

FIG. 3. A segment of lampbrush chromosome from an oocyte of the newt, *Triturus*. Pairs of lateral loops project from an axis of Feulgen-positive granules.

FIG. 4. Postulated structure of a pair of lampbrush chromosome loops. The loop consists of ribonucleoprotein matrix surrounding a very delicate DNA fibril.

activity. Since the loops are structurally simpler than the puffs, the problem is obviously easier to attack in the lampbrush chromosomes. Several years ago Callan and Macgregor (1958) showed that lampbrush chromosome loops, although Feulgen-negative, are extensively fragmented by deoxyribonuclease (DNase) (Figs. 5–9). This observation is of great interest in showing that DNA occurs in the loops and indeed is involved in maintaining structural continuity of the loop. The observation by itself does not prove that DNA forms a continuous fiber throughout the length of the loop; however, it is significant that ribonuclease (RNase) and several proteases do not cause fragmentation (Macgregor and Callan, 1962). It seems possible that the looped fibers which traverse the puffs in the polytene chromosomes also contain DNA, since the puff forms at the "expense" of a DNA band, and a DNA band is reconstituted when the puff regresses (Beermann, 1952a).

Callan and Macgregor (1958) also noticed that the delicate fiber connecting successive chromomeres in the lampbrush chromosome is attacked by DNase. Consequently, the enzyme not only fragments the loops but also reduces the chromosome to a number of shorter segments (Figs. 5 and 6). If the interpretation given in Fig. 4 is correct in its essentials, we should expect that breakage of the interchromomeric fiber would involve enzymatic attack on two chromatids, whereas breakage of a loop would involve only one chromatid.

Recently an effort has been made to clarify these points by making a kinetic analysis of the DNase fragmentation (Gall, 1963). The lampbrush chromosomes provide excellent material for such a study since they can be isolated unfixed into the enzyme solution. As the material fragments, one can count the pieces and analyze the number of breaks as a function of time. If we assume that the chromosome consists of n longitudinal subunits which are attacked independently and with equal probability by the enzyme, then breakage should follow the equation:

$$b = k_1 t^n \tag{1}$$

in which b is the observed number of breaks, k_1 is a proportionality constant, and t is time. The determination of n, the number of subunits, is made by plotting the data in the form:

$$\log b = n \log t + k_2 \tag{2}$$

Here n will be the slope of the line obtained by plotting $\log b$ against $\log t$. It was found that n for the loops averaged 2.6 (23 determinations), whereas for the interchromomeric regions the value was 4.8 (19 determinations). The simplest interpretation of these data would seem to be that there are two independently attacked subunits in the loop, but four in the interchromomeric regions. The data are certainly not consistent with the assumption of 1 and 2 or 4 and 8 subunits in these regions, respectively, although we have no experimental reason to rule out the possibility of odd numbers or non-integers.

The nature of the subunits cannot be deduced from the breakage kinetics, but only from what is known of the specificity of the enzyme. The enzyme used, DNase 1 from beef pancreas, probably breaks the two polynucleotide chains of the DNA molecule independently, as shown by kinetic analysis of DNA degradation in solution (Schumaker *et al.*, 1956; Thomas, 1956). If the enzyme behaves in the same fashion when digesting the DNA of the chromosome, then the data suggest that the DNA of the chromatid exists as one double helix. Our information does not distinguish between one very long molecule in the loop or a series of shorter molecules connected end to end. It is possible, of course, that orientation of the DNA in the chromosomes is such as to insure preferential attacks by the enzyme, in which case the observed value of *n* would be smaller than the true number of polynucleotide chains.

The number of polynucleotide strands cannot be very large, however, since the loop axis—the delicate thread on which the ribonucleoprotein matrix is accumulated—is no more than 60–80 Å in diameter (Figs. 10 and 11). The double helix of the DNA molecule is about 20 Å in diameter and is presumably associated with protein in the chromosome.

It is much more difficult to make statements about the thousand or more threads of the polytene chromosomes. Heitz and Bauer (1933) long ago noticed that the interband regions of the salivary gland chromosomes are faintly Feulgen-positive. Recently Swift (1962) has made an attempt to correlate the intensity of Feulgen staining in the interband regions with the intensity expected for various degrees of DNA strandedness. He used the large chromosomes of *Sciara* and the Azure A-Feulgen technique, which gives considerably darker staining than the conventional Feulgen. On the assumption that continuous DNA fibers span the interband regions, and knowing the level of polyteny in the cells examined, he concluded that there could be only one or two DNA double helices in each of the chromosome fibers. Since we do not know that the DNA is continuous we must accept these results with caution. They do render unlikely any model of the chromosome which postulates large numbers of polynucleotide chains in each chromatid (cf. also the discussion in Taylor, 1963).

II. Loops, Puffs, and Genes

Extensive studies during the past decade have shown that puffing of the chromosome bands is a highly specific phenomenon, a given tissue being characterized by unique patterns of puffing during the course of its develop-

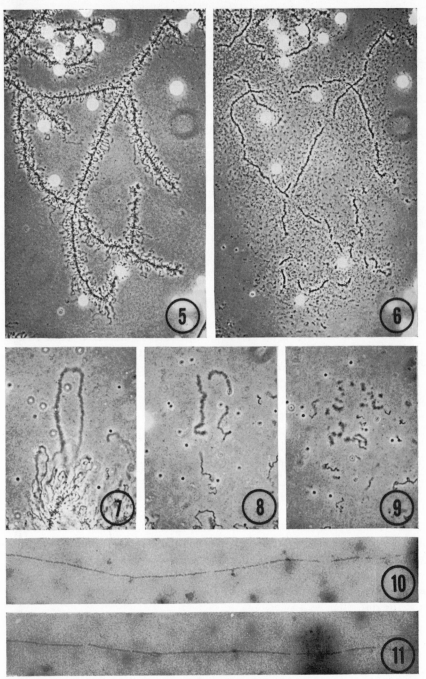

FIG. 5. A lampbrush chromosome (bivalent), unfixed, 10 minutes after being isolated in a DNase solution (0.02 mg/ml). Phase contrast, 200×.

FIG. 6. The same chromosome, after 30 minutes in DNase. Note that the loops have been extensively fragmented, and that the main chromosome axis is broken in a number of places. Phase contrast, 200×.

ment (Beermann, 1952a,b, 1956; Breuer and Pavan, 1953, 1955; Mechelke, 1953; Becker, 1959; Kroeger, 1960). The chromosomes of *Chironomus, Sciara,* and their close relatives have been favorite objects of study because of their large size, but the same conclusions hold for *Drosophila* and probably for other genera as well.

Beermann has argued that the time- and tissue-specific puffing of certain bands should be interpreted as cytological evidence of differential gene activation. A summary statement of his views may be found in an earlier volume of the Growth Symposia (Beermann, 1959). Recently he has described a case in which the correlation between puffing and gene activity is quite clear (Beermann, 1961). In *Chironomus tentans* the salivary secretion lacks a special kind of granule which is found in *Chironomus pallidivittatus* and in all other species examined. These granules are produced by four special cells near the duct of the salivary gland (Fig. 12). Crosses between *C. tentans* and *C. pallidivittatus* show that the inability to produce the granules (SZ granules, named after their production in "Sonderzellen") is inherited as a simple recessive Mendelian character. The locus of the *sz* gene was shown to be in chromosome IV, either in its distal region or directly adjacent to the centromere. The localization was done by conventional cytogenetic techniques making use of inversions as markers. Immediately next to the centromere, that is, in one of the two possible locations of the gene, a conspicuous puff (Balbiani ring) is found, but only in the four special gland cells of *C. pallidivittatus* which produce the SZ granules. The puff is absent from the rest of the salivary gland cells in *C. pallidivittatus* and from all the cells of *C. tentans* (Fig. 13). In hybrid individuals the puff is "heterozygous"; that is, only the chromosome carrying the dominant allele, *sz*+, shows a puff. These observations constitute the most compelling evidence now available that puffing is a visible manifestation of gene activity.

As Beermann has pointed out, it would be of the greatest interest to know just what biochemical steps lead to the production of the special salivary secretion and how many of these steps are controlled by the *sz* locus. Conceivably the *sz* locus is concerned with the production of only one protein, which is needed in relatively large amounts by the cells producing the secretion. On the other hand, the locus could be complex and involve a whole

---

FIGS. 7–9. Three photographs of the same loop taken 5, 13, and 16 minutes after isolation in DNase (0.02 mg/ml). Fragmentation of the loops occurs without noticeable decrease in width of the pieces. Phase contrast, 300×.

FIGS. 10 AND 11. Electron micrographs of the fibril which forms the axis of the loop and which is responsible for maintaining structural continuity of the loop. It is presumed that this fibril consists of DNA and associated protein; its diameter is about 60–80 Å. 75,000×. (Micrographs furnished by Dr. O. L. Miller, Oak Ridge National Laboratory.)

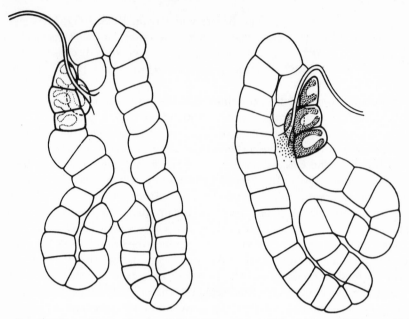

FIG. 12. The salivary glands of *Chironomus tentans* (*left*) and *C. pallidivittatus* (*right*) Note the four special cells ("Sonderzellen") in *C. pallidivittatus*, which produce a granular secretion. The production of secretion is controlled by the sz gene located in chromosome IV. Individuals lacking the secretion are sz/sz, while those producing secretion are sz^+/sz^+ or sz^+/sz. From Beermann (1961).

series of ordered reactions. Further genetic analysis will be difficult because of the lack of suitable allelic forms, but valuable information can probably be obtained from chemical studies on the salivary secretion itself.

It has not yet been possible to correlate the appearance of a particular loop pair in the lampbrush chromosomes with a known gene. Certain attributes of the loops, however, parallel those of the "classical" gene, and these may be briefly mentioned. On a chromosome of average length in *Triturus* there are of the order of 1000 loop pairs. Despite their long physical length, therefore, each corresponds to a relatively short segment of the genetic material. The loops are morphologically heterogeneous, and they have individually distinctive rates as well as patterns of precursor uptake (see Section IV). Callan and Lloyd (1956, 1960a,b) have shown that loops at a given position on the chromosome can exist in alternative forms distinguishable by specific morphological features. A given individual can be homozygous or heterozygous for a particular loop morphology and individual peculiarities of loop morphology are transmitted in a regular Mendelian

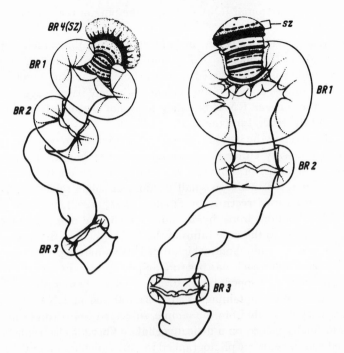

Fɪɢ. 13. *Left:* chromosome IV from a special cell of an individual capable of producing the secretion (sz^+/sz^-). Note the puff labeled BR 4(SZ) at the tip of the chromosome. *Right:* chromosome IV from the same individual (sz^-/sz^+), but from one of the normal cells which does not produce the granular secretion. The tip of the chromosome shows normal banding. From Beermann (1961).

fashion to the offspring. Callan and Lloyd (1960a,b) have also shown that within a natural population of newts the relative numbers of individuals homozygous or heteroxygous for particular loop forms are consistent with the Hardy-Weinberg relationship.

III. Heterogeneity of Nuclear RNA

The idea that RNA is concerned in the transfer of the genetic code from the DNA of the chromosomes is a major tenet in nearly all recent theories of gene action (see various papers in Frisch, 1961). In general at least three types of RNA are recognized: soluble or transfer RNA, of relatively low molecular weight and specific for the different amino acids; messenger RNA, distinguished by high metabolic rate and by possessing base ratios similar to the DNA of which it is presumably a copy; and ribosomal RNA, whose role is still unclear. Ribosomal RNA makes up the bulk of the cytoplasmic RNA on a weight basis.

Considerable evidence is now at hand to indicate that much of the cell's RNA is made in the nucleus. No attempt will be made here to review the literature, which is well covered in the article of Prescott (1960). The incorporation studies, on which this evidence is largely based, suggest a heterogeneity of the nuclear RNA, but they have not permitted precise characterization of the several fractions.

If we are to relate the various nuclear RNA fractions to specific cytological areas, it is necessary either to use bulk isolation techniques for the components of the nucleus, or, conversely, so to refine the analytical techniques that studies may be made on small portions of single nuclei. The first approach has been used recently by Sibatani *et al.* (1962), who have isolated several RNA fractions from thymus nuclei, including one whose base ratios are rather similar to the base ratios of the DNA (making the usual substitution of uracil for thymine) and which they believe may represent the messenger. The second approach has been made possible by the elegant technique of microphoresis developed by Edström (1960a). Base ratio analyses are possible on samples containing as little as 100 $\mu\mu$g of RNA. The RNA is extracted enzymatically from the sample, subjected to hydrolysis in a microdrop, and finally placed on a delicate cellulose fiber for electrophoresis. The separated components are photographed in ultraviolet light and their amount estimated photometrically. The sample may be a small bit of material collected with a microneedle from a fixed tissue slice or squash.

A. *Nucleoli*

Base ratio analyses of nucleolar RNA have been made by Edström and his co-workers on material from oocytes of a spider (Edström, 1960b), a starfish (Edström *et al.*, 1961), and two species of newt (Edström and Gall, 1963), and on the nucleoli from *Chironomus* salivary gland cells (Edström and Beermann, 1962). These RNA samples differ conspicuously from one another, as shown in Table I. An extremely interesting relationship emerges, however, when one compares the cytoplasmic RNA from the same cells, also shown in Table I. In each case the cytoplasmic RNA tends to have a composition similar to the RNA of the nucleolus. This can be seen by plotting the molar proportion of each base in the cytoplasm against the molar proportion of the same base in the nucleolus (Fig. 14). The RNA analyzed in these cases is simply that which remains in the nucleolus or cytoplasm after fixation in Carnoy's fluid or formaldehyde. The cytoplasmic RNA represents for the most part ribosomal RNA. The original papers should be consulted for a cautious consideration of the errors inherent in the technique.

TABLE I

Base Composition of RNA from Nucleolus and Cytoplasm of Five Animal Species[a],[b]

	Tegenaria (Spider) Oocyte		Asterias (Starfish) Oocyte		Chironomus (Fly) Salivary gland		Triturus cristatus (Newt) Oocyte		Triturus viridescens (Newt) Oocyte	
	N[c]	C[c]	N	C	N	C	N	C	N	C
Adenine	25.2	25.1	23.7	23.5	30.6	29.4	18.1	20.1	21.5	23.7
Guanine	29.8	30.2	33.4	31.9	20.1	22.9	31.7	27.2	29.3	28.0
Cytosine	22.9	21.9	24.3	24.8	22.1	22.1	28.7	29.5	30.1	27.7
Uracil	22.2	22.9	18.5	19.7	27.1	25.7	21.7	23.0	19.1	20.7

[a] Mean values of molar proportions in per cent of the sum.
[b] Data from Edström and co-workers (Edström, 1960a,b; Edström and Beerman, 1962; Edström and Gall, 1963; Edström et al., 1961).
[c] N = nucleolus; C = cytoplasm.

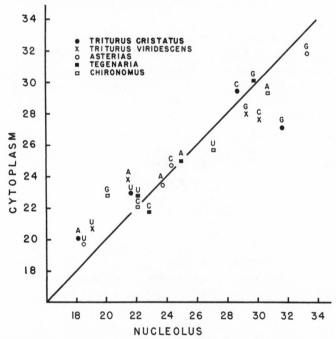

Fig. 14. Graph to illustrate the similarity in composition of cytoplasmic and nucleolar RNA in five different animal species. The scale on the axes gives the molar proportion of purine or pyrimidine in per cent of the sum; each point represents one base in one species. (A,G,C,U = adenine, guanine, cytosine, uracil). If the cytoplasmic and nucleolar RNA were identical, points would fall along the straight line. Data from Edström and co-workers (see Table I).

TABLE II

BASE COMPOSITION OF CHROMOSOMAL RNA FROM OOCYTES OF THE NEWT
Triturus cristatus AND SALIVARY GLAND CELLS OF *Chironomus tentans*[a,b]

	Triturus[c]	Chironomus			
	All chromosomes together	Chromosome I	Chromosome IV		
			Upper	Middle	Lower
Adenine	26.3	29.4	35.7	38.0	31.2
Guanine	20.7	19.8	20.6	20.5	22.0
Cytosine	24.7	27.7	23.2	24.5	25.4
Uracil	28.3	23.1	20.8	17.1	20.2

[a] Excludes nucleolar and nuclear sap RNA. Expressed as molar proportions in per cent of the sum.
[b] Data from Edström and Beermann (1962) and Edström and Gall (1963).
[c] In *Triturus* the DNA composition is: A = T = 28.8; G = C = 21.2.

B. *Chromosomes*

Although nucleoli usually represent the most obvious accumulation of RNA in the nucleus, RNA is also found in the nuclear sap and the chromosomes. In the amphibian oocyte, for instance, the chromosomes and sap together contain more RNA than the nucleoli. Non-nucleolar RNA similarly makes up a large fraction in the other cells analyzed by Edström. Base ratios have been determined for pooled chromosomal RNA from oocytes of two species of newt and for individual chromosomes and parts of chromosomes of *Chironomus*. The results of these analyses are shown in Table II. In the two species of *Triturus* it is clear that the pooled chromosomal RNA is quite unlike the RNA of the nucleoli or cytoplasm. However, the base ratios are rather similar to the DNA base ratios of the same species, substituting uracil for thymine. The DNA base ratios were determined from red blood cell nuclei.

In the salivary glands of *Chironomus*, Edström and Beermann (1962) have made a direct demonstration of RNA heterogeneity between parts of the same chromosome. The small chromosome IV of *Chironomus tentans* has three conspicuous puffs which together account for a large fraction of the total RNA of this chromosome (Fig. 13). By collecting several hundred chromosomes and cutting them into three pieces, Edström and Beermann have been able to analyze the RNA of relatively small parts of the genome. They also analyzed the total RNA of chromosome I. The RNA extracted from chromosome I differed significantly from that found in the segments of chromosome IV, and the latter differed significantly between each other (Table II).

Any generalizations to be made from these data must be tentative because of the small number of cases examined. It seems probable that the chromosomal RNA is heterogeneous, as shown by the analyses between and within chromosomes. These differences might be referable to heterogeneity at the individual locus level, although as yet the smallest fraction analyzed represents about ⅓ of the smallest chromosome of *Chironomus*. The pooled chromosomal RNA in *Triturus* resembles the DNA more closely than it does other RNA fractions. Nevertheless the two differ significantly and it is not possible to say that the one is a complete copy of the other. The cytoplasmic RNA does not resemble the chromosomal RNA at all closely. It must, therefore, represent either a very small fraction of the chromosomal RNA, or something else entirely. Because of its close similarity to nucleolar RNA we may postulate that the cytoplasmic RNA (ribosomal RNA) is primarily a product of the nucleolus.

The nucleolus, as a conspicuous and nearly universal component of cells, has received considerable attention in all discussions of RNA metabolism (see reviews of Vincent, 1955; Swift, 1958). Basically two ideas have been entertained about its function: that it represents a storehouse into which material from all parts of the chromosomes is dumped before being transmitted to the cytoplasm, or that it represents the specific product of one genetic locus. The first of these, as attractive as it is on many grounds, is difficult to reconcile with the base ratio analyses. Nor is it supported by certain tracer studies, which seem to show an independent metabolic pathway for RNA synthesis in the nucleolus (McMaster-Kaye and Taylor, 1958; McMaster-Kaye, 1962). The postulate that the primary function of the nucleolus is the production of ribosomes fits the base ratio analyses, which show a close correspondence between nucleolar and cytoplasmic RNA; and also agrees with the well known observation that nucleoli contain small particles which are morphologically similar to the ribosomes (Fig. 15). It also helps to explain why the nucleolus should be of very nearly universal occurrence in animal and plant cells, and why its elimination leads ultimately to death (McClintock, 1934; Elsdale *et al.*, 1958; Beermann, 1960).

According to current notions the ribosomes must become "loaded" with messenger RNA before they are functional in protein synthesis. If the ribosomes are produced in the nucleolus, then we must ask where and under what circumstances they become associated with the chromosomal RNA. Conceivably this association could take place before the ribosomes leave the nucleolus for the cytoplasm, or it could occur in the cytoplasm, both ribosomes and chromosomal RNA traveling independent paths out of the nucleus. Some of the conflicting evidence regarding precursor pathways in the nucleus (Harris, 1959; Fitzgerald and Vinijchaikul, 1959; Amano and Leblond, 1960;

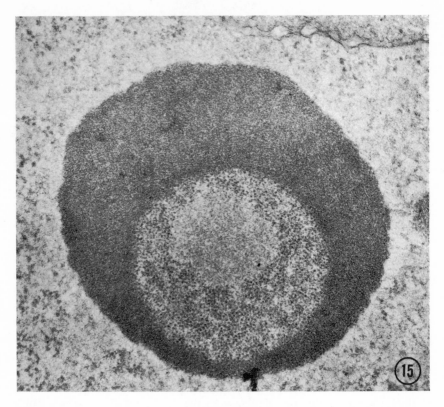

FIG. 15. Electron micrograph of a nucleolus from the oocyte of *Triturus viridescens*. A dense peripheral area surrounds a central cluster of ribosomelike particles. Such particles are found in a wide variety of nucleoli. Section of OsO_4-fixed oocyte stained with UO_2. 30,000×. (Micrograph furnished by Dr. O. L. Miller, Oak Ridge National Laboratory. See also Miller, 1962.)

McMaster-Kaye, 1962) might be resolved if the nucleolus at times is actively engaged in the synthesis of new RNA (ribosomal) and at other times is receiving RNA from the chromosomes. Such a view would be a compromise between the "storehouse" and "gene-product" theories of the nucleolus.

IV. RNA Synthesis in Lampbrush Chromosome Loops

Morphological studies on the lampbrush chromosomes, culminating in the important monograph of Callan and Lloyd (1960b), have shown that the loops come in a wide variety of forms. All have one feature in common: one end of the loop is an extremely thin thread, often barely discernible by phase contrast microscopy, whereas the rest of the loop is broader (Fig. 3). Some loops may be gourdlike with very marked asymmetry, while others are more slender and tapering. The majority of loops are of nearly uniform diameter, 1 or 2 μ throughout most of their length. Added to these differences in asymmetry are marked differences in the texture of the ribonucleoprotein making up the bulk of the loop. Hence individual loops may be easily recognized and mapped.

Several years ago it was suggested (Callan, 1956; Gall, 1955) that loop asymmetry might be explained on two assumptions: (1) the loop is formed by the spinning out of a thread from the parent chromomere, and (2) synthesis of ribonucleoprotein occurs on this thread, but is not accompanied by immediate release of the product. On these assumptions the loop would be spinning out at the thin end, which had consequently had less time to accumulate ribonucleoprotein matrix than the older, thick end.

We have recently tried to test this hypothesis by following the incorporation of H^3-uridine in the chromosomes of two species of newt, *Triturus cristatus* and *T. viridescens* (Gall and Callan, 1962, and unpublished observations). Chromosomes may be labeled either by injecting the radioactive compound directly into the animal or by placing bits of the ovary into a dish containing the isotope. The second method makes it easy to obtain high activity in the chromosomes and to study short term incorporation. Within an hour or two after isotope administration, the majority of loops become labeled along their whole length. There are marked variations in intensity of label from one loop to the next, suggesting a heterogeneity of metabolic rates (Fig. 16). A very few loops remain unlabeled after such brief exposure, and these have proved of considerable interest.

Two such nonlabeling loops have been examined in detail, one in each of the two species. When H^3-uridine is injected into the coelomic cavity, these loops begin to show labeling after about a day, but the radioactivity in every case is limited to a short region near the *thin* end of the loop. During the course of about 10 days the area of labeling extends from the thin end toward the thick, so that eventually the whole loop is labeled (Fig. 17).

Labeling with H^3-phenylalanine follows a different course. Here all loops, including the ones which label sequentially with uridine, are labeled along their entire length after 1 day.

In the experiment with uridine one must distinguish between an actual movement of materials from one side of the loop to the other, and a wave of synthesis starting at the thin end and passing over the loop. Theoretically, one should be able to distinguish these alternatives by diluting out the precursor at a suitable time and noting the change in the pattern of incorpora-

tion. Unfortunately, we have not been able to affect the pattern of incorporation either by injecting nonradioactive uridine or by transferring a labeled
ovary to a cold animal. Nevertheless, it seems unlikely that we are dealing
with a wave of synthesis, since all cases showed labeling first at the thin end
of the loop. If waves of synthesis were passing over the loops, we should have
found oocytes in different phases at the time of isotope administration. That
is, in some cases the initial labeling should have appeared at intermediate
points along the loop.

For the sake of argument, let us accept that sequential labeling results from
a movement of material from one side of the loop to the other. There are
still two alternatives: the DNA loop axis is stationary while the ribonucleoprotein matrix moves along it, or the axis and matrix move together like a
conveyor belt. Both models would result in the observed sequential labeling.
In the first case only a very short region of the DNA fiber would be concerned
(continuously) in the production of RNA. In the second model, the loop
would be spinning out all the time, each new segment of fiber being engaged
in RNA synthesis for a short time. The second hypothesis requires the sub-
sidiary assumption that the loop axis is reeling in at the thick end of the loop
and that material is being shed from the thicker regions of the loop. We have
had circumstantial evidence for a long time that material is shed from the
loops (Duryee, 1950; Gall, 1955; Callan, 1956). It is also clear that the loop
eventually spins back in during the later stages of oogenesis, when all the
loops regress; and the loop axis contracts into the chromomere when the
ribonucleoprotein matrix is removed experimentally (Gall, 1956; Macgregor
and Callan, 1962). Hence, it seems reasonable to suppose that the loop could
be spinning out continuously on one side and back in on the other side.

On either model (stationary or moving loop axis) the asymmetry of the
loop would be explained by the continuing protein synthesis. As already
mentioned, phenylalanine incorporation occurs throughout the length of the
loop. Presumably then, the matrix material synthesizes protein during its
10-day journey from one side of the loop to the other.

What are we to say of the majority of loops which do not show sequential
uridine labeling? Obviously uridine incorporation in these loops is not
localized and we cannot use its translocation as a measure of loop movement.
However, it does not follow that the majority of loops are stationary. On the
contrary, the asymmetry of the loops itself suggests that movement is the
usual situation, the thicker end of the loops being the older. In these cases
we have to postulate that both RNA and protein synthesis go on continuously
as the material moves along. We believe that what is special about the loops
with sequential labeling is not that they move while others are stationary,

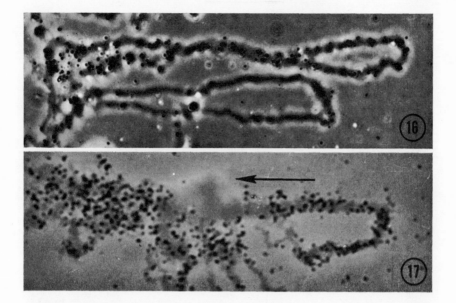

FIG. 16. Autoradiograph of two adjacent loops on the same chromosome showing marked difference in isotope incorporation. From oocyte of *Triturus* injected 26 hours previously with H³-uridine. 1500×.

FIG. 17. The large loop on chromosome I of *Triturus viridescens* which labels sequentially with H³-uridine. Autoradiograph of the loop and adjacent chromosome region from an individual injected 3 days previously. The labeling began at the thin end of the loop (below right) and is proceeding in the direction of the arrow. The thick end of the loop is the large mass of unlabeled material in the center of the photograph. About 1500×.

but rather that their movement is detectable because of unusually restricted RNA synthesis.

V. Control of Chromosomal Synthesis

During larval development in *Chironomus* or *Drosophila* the chromosomes of a given organ go through a regular sequence of puffing and regression at specific loci. The lampbrush chromosomes, too, exhibit changes during oocyte enlargement, as carefully detailed by Callan and Lloyd (1960b). Certain conspicuous loops which are present in the later stages of oocyte development are not evident in earlier stages; and other loops slowly change their morphology during the long period of oocyte enlargement. If these changes do indeed represent differential activity of genes or groups of genes, then study of the factors affecting puff and loop formation might well provide information on genetic control mechanisms.

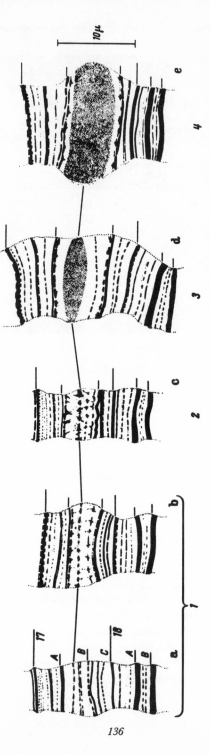

Fig. 18. Various degrees of puffing at band I-17-B in *Chironomus*; *a–c* from untreated control animals, *d* and *e* from larvae which had been injected with the hormone ecdysone. From Clever (1961).

In his study of salivary gland puffs of *Drosophila* Becker (1959) showed that glands isolated into a salt solution may continue their normal sequence of puff formation and regression for at least an hour. In this case the changes which occur may well be ones which were "triggered" before the gland was isolated, and the relative autonomy of the system be related to the short duration of the experiment. The observation is important in showing that experimental manipulation of isolated organs is possible. It would be interesting to know, for instance, if various simple compounds added to the external medium could cause recognizable new puffs or regression of old ones.

External factors, on the other hand, certainly do influence puffing and loop formation. The most thorough study of the control of puffing has been carried out by Clever (1961), who has examined the effects of the insect hormone ecdysone on the salivary gland puffs of *Chironomus* (Fig. 18). When a small amount of the hormone is injected into a larva of the proper stage, pupation occurs within a few days, even though the animal would not ordinarily pupate for a week or more. Externally no morphological changes are evident for some time after the injection. Clever has shown, however, that the hormone has an almost instantaneous effect on the puffing pattern of the chromosomes. Within 15–30 minutes after injection of the hormone a new puff appears at locus I-18-C. This is followed within about an hour by a new puff at locus IV-2-B. Eventually a whole series of puffs occur in a regular order. The sequence of puffing which precedes the induced molting is similar to the sequence which occurs at the time of normal pupation. Clever postulates that the hormone has its primary effect on the gene at locus I-18-C, and this locus in turn activates still other genes in a chain reaction. In this way a relatively simple external influence may trigger a complex series of intranuclear reactions leading eventually to the physiological changes associated with molting.

In terms of the schemes discussed by Jacob and Monod (1961), the hormone might act by combining with a repressor inside the cell. This would lead in turn to the derepression of a given gene, in this case I-18-C. Further activation of genes would result either from the direct action of the ribonucleoprotein product of gene I-18-C or by the action of smaller molecules produced under its influence. It would appear that the salivary gland chromosomes offer exceptionally good experimental material for further studies on regulator mechanisms in cells of higher organisms.

Two other cases of altered chromosome activity may be mentioned. Kroeger (1960) has transplanted nuclei from larval salivary glands of *Drosophila* into the cytoplasm of the egg. He has reported that certain puffs regress whereas other new ones arise, the effects being dependent upon the developmental stage of the eggs used.

Macgregor (1963) has followed the changes in lampbrush chromosomes from oocytes of *Triturus cristatus* after injection of gonadotropic hormone. Several quite striking alterations in loop morphology occur, as well as changes in the consistency of the nuclear sap. In this instance, no "target" chromosome locus has been discovered comparable to band I-18-C in Clever's study. Another intriguing effect of gonadotropic hormone is to speed up the rate of H^3-uridine incorporation in the sequentially labeled loops (Gall and Callan, 1962). Whereas total labeling occurs after about 10 days in normal animals, those given the hormone show completely labeled loops after four days.

VI. Discussion

The major concern of this paper has been to define, in morphological and cytochemical terms, the nature of the units in the chromosome concerned with RNA synthesis. These units are the loops of the lampbrush chromosome and the puffs of the polytene chromosome. One of their most striking features is their size. The threads which make up the larger puffs or Balbiani rings are some 5 μ or more in length; the loops of the lampbrush chromosomes average perhaps 50 μ, exceptional cases reaching 200 μ or more. The DNA, or DNA-protein fiber which maintains structural continuity in a lampbrush chromosome loop, is, however, less than 100 Å in diameter, and may contain only one Watson-Crick double helix.

A DNA fiber of these dimensions is not what the geneticist usually has in mind when he speaks of a gene, in the sense of a cistron. Benzer (1961) has estimated that the A cistron in the *rII* region of phage T4 may consist of several hundred nucleotide pairs. A DNA fiber 1000 nucleotide pairs in length (sufficient to specify a protein of 333 amino acids on the basis of a triplet code), would be slightly more than 0.3 μ in length.

There are several ways to reconcile the large size of the chromosome units with the suggested dimensions of the cistron: (1) The units we are dealing with in the chromosome are complex, perhaps consisting of an integrated series of different cistrons; (2) the chromosome units are highly repetitive, there being many copies of the same cistron; (3) there is a great deal of DNA not concerned with coding but in some unknown way essential to the activity of the unit; or (4) the DNA segments are short and interspersed with long segments of protein or other material.

We can rule out the last of these alternatives on the grounds that DNase reduces the loops to extremely short segments, most of which are on the borderline of visibility in the light microscope or smaller. This observation does not necessarily imply that the DNA is continuous, but it does tell us that there are no very long segments which are free of DNA.

Among the other alternatives we cannot at the moment make any certain decision and the actual situation may involve a combination of the choices. The extraordinary length of the lampbrush chromosome loops in *Triturus* is probably correlated with another peculiarity of salamanders. This is the exceptionally high DNA content of their diploid cells, first recognized in the DNA measurements of Mirsky and Ris (1951). The content per diploid nucleus in various urodeles ranges from about 7 to 30 times that in a typical mammal such as the mouse or rat (Mirsky and Ris, 1951; Gall, 1962). This fact is usually explained by assuming that the chromosomes of urodeles have a higher degree of polyteny than the chromosomes of mammals; that is, they consist of more strands, but the strands themselves are similar in length. The DNase experiments render this interpretation unlikely. More probably the chromosome length varies from organism to organism and approximates or equals the total length of the contained DNA. The very high DNA content of the urodeles might have been derived by serial replication of units along the chromosome strand, rather than by piling up of a large number of strands. Such a process, with its consequent redundancy of genetic information, could have led to the great length of the lampbrush chromosome loops. In the course of evolution the various replicated subunits might diverge by mutation and lead to greater genetic variability.

Whatever may be the fine structure of the units with which we are dealing, we do know that they are involved in RNA synthesis and that their periods of synthetic activity can be correlated, as shown so elegantly by Beermann and others, with functional changes in the cell.

Most of the cytochemical problems of chromosomal RNA synthesis lie ahead of us. Can it be shown, for instance, that the RNA made at each point on the chromosome is unique and specific for that point? If the microphoresis technique of Edström could be improved in sensitivity by a factor of 10, or better 100, perhaps by combining it with radioisotopes, then it should be possible to obtain base ratio analyses on single loops and the smaller puffs.

In the discussion of the RNA of the chromosome, we should not lose sight of the fact that the bulk of the material on the loops and puffs is protein, most of which may be synthesized in place. Is this some kind of "carrier" protein necessary for the transport of RNA to the nucleolus or to the cytoplasm? Or is it a more specific product of the locus itself? Conceivably this question could be approached through the use of the fluorescent antibody technique.

If the DNA of the loop is laid out in single file, as we now believe, it should be possible to study the type of connection between molecules, the state of the molecule at the time of RNA synthesis, and similar problems. A number of enzymes are now available which might give insight into such structural questions. Finally, if it becomes possible to obtain RNA synthesis by the chromosomes after isolation from the nucleus, the way would be open for a wide variety of studies relating to the mechanism of gene action.

All these approaches would be of general interest to problems of differentiation, since the control of gene action appears to be closely linked with cellular differentiation. At present we know only one agent, the insect hormone ecdysone, which has a specific regulatory action on a chromosome puff. Hopefully, other such substances will be found and eventually help us to identify the nature of the intranuclear control mechanisms.

VII. Concluding Remarks

The purpose of this paper has been to make the following statements seem reasonable. They are intended more as hypotheses for future work than as rigorous conclusions from available evidence.

a. The chromosomes of higher organisms are composed of one, or at most a few, DNA double helices that may extend the length of the chromatid.

b. The functional units, in terms of RNA synthesis, are DNA segments up to 50 μ or more in length (the loops of the lampbrush chromosomes and the puffs of the polytene chromosomes). These units must consist of many cistrons.

c. The units are polarized. RNA synthesis begins at one end of the unit and proceeds sequentially to the other.

d. The chromosomal RNA is heterogeneous. It is probably equivalent to the messenger RNA of bacteria.

e. The ribosomal RNA is produced in the nucleolus.

f. Cellular differentiation is associated with differential activation of the chromosomal units.

REFERENCES

AMANO, M., AND LEBLOND, C. P. (1960). Comparison of the specific activity time curves of ribonucleic acid in chromatin, nucleolus and cytoplasm. *Exptl. Cell Research* **20**, 250–253.

BECKER, H. J. (1959). Die Puffs der Speicheldrüsenchromosomen von *Drosophilia melanogaster. Chromosoma* **10**, 654–678.

BEERMANN, W. (1952a). Chromomerenkonstanz und spezifische Modifikationen der Chromosomenstruktur in der Entwicklung und Organdifferenzierung von *Chironomus tentans. Chromosoma* **5**, 139–198.

BEERMANN, W. (1952b). Chromosomenstruktur und Zelldifferenzierung in der Speicheldrüse von *Trichocladius vitripennis. Z. Naturforsch.* **7b**, 237–242.

BEERMANN, W. (1956). Nuclear differentiation and functional morphology of chromosomes. *Cold Spring Harbor Symposia Quant. Biol.* **21**, 217–232.

BEERMANN, W. (1959). Chromosomal differentiation in insects. *In* "Developmental Cytology," Growth Symposium No. 16 (D. Rudnick, ed.), pp. 83–103. Ronald Press, New York.

BEERMANN, W. (1960). Der Nukleolus als lebenswichtiger Bestandteil des Zellkernes. *Chromosoma* 11, 263–296.

BEERMANN, W. (1961). Ein Balbiani-Ring als Locus einer Speicheldrüsenmutation. *Chromosoma* 12, 1–25.

BENZER, S. (1961). On the topography of the genetic fine structure. *Proc. Natl. Acad. Sci. U.S.* 47, 403–415.

BREUER, M. E., AND PAVAN, C. (1954). Salivary chromosomes and differentiation. *Proc. 9th Intern. Congr. Genetics, Florence, Italy, 1954* (Suppl. of *Caryologia* 6), Part II, p. 778.

BREUER, M. E., AND PAVAN, C. (1955). Behavior of polytene chromosomes of *Rhynchosciara angelae* at different stages of larval development. *Chromosoma* 7, 371–386.

CALLAN, H. G. (1956). Recent work on the structure of cell nuclei. In Symposium on the Fine Structure of Cells, Leiden. *Intern. Union Biol. Sci. Publ.* B21, 89–109.

CALLAN, H. G. (1962). The nature of lampbrush chromosomes. *Intern. Rev. Cytol.* 15, in press.

CALLAN, H. G., AND LLOYD, L. (1956). Visual demonstration of allelic differences within cell nuclei. *Nature* 178, 355–357.

CALLAN, H. G., AND LLOYD, L. (1960a). Lampbrush chromosomes. *In* "New Approaches in Cell Biology" (P. M. B. Walker, ed.), pp. 23–46. Academic Press, New York.

CALLAN, H. G., AND LLOYD, L. (1960b). Lampbrush chromosomes of crested newts *Triturus cristatus* (Laurenti). *Phil. Trans. Roy. Soc.* B243, 135–219.

CALLAN, H. G., AND MACGREGOR, H. C. (1958). Action of deoxyribonuclease on lampbrush chromosomes. *Nature* 181, 1479–1480.

CLEVER, U. (1961). Genaktivitäten in den Riesenchromosomen von *Chironomus tentans* und ihre Beziehungen zur Entwicklung. 1. Genaktivierung durch Ecdyson. *Chromosoma* 12, 607–675.

DURYEE, W. R. (1950). Chromosomal physiology in relation to nuclear structure. *Ann. N.Y. Acad. Sci.* 50, 920–953.

EDSTRÖM, J.-E. (1960a). Extraction, hydrolysis, and electrophoretic analysis of ribonucleic acid from microscopic tissue units (microphoresis). *J. Biophys. Biochem. Cytol.* 8, 39–46.

EDSTRÖM, J.-E. (1960b). Composition of ribonucleic acid from various parts of spider oocytes. *J. Biophys. Biochem. Cytol.* 8, 47–51.

EDSTRÖM, J.-E., AND BEERMANN, W. (1962). The base composition of nucleic acids in chromosomes, puffs, nucleoli, and cytoplasm of *Chironomus* salivary gland cells. *J. Cell Biol.* 14, 371–380.

EDSTRÖM, J.-E., AND GALL, J. G. (1963). The base composition of ribonucleic acid in lampbrush chromosomes, nucleoli, nuclear sap, and cytoplasm of *Triturus* oocytes. In preparation.

EDSTRÖM, J.-E., GRAMPP, E., AND SCHOR, N. (1961). The intracellular distribution and heterogeneity of ribonucleic acid in starfish oocytes. *J. Biophys. Biochem. Cytol.* 11, 549–557.

ELSDALE, T. R., FISCHBERG, M., AND SMITH, S. (1958). A mutation that reduces nucleolar number in *Xenopus laevis*. *Exptl. Cell Research* 14, 642–643.

FITZGERALD, P. J., AND VINIJCHAIKUL, K. (1959). Nucleic acid metabolism of pancreatic cells as revealed by cytidine-H³ and thymidine-H³. *Lab. Invest.* 8, 319–328.

FRISCH, L., ed. (1961). *Cold Spring Harbor Symposia Quant. Biol.* 26.

GALL, J. G. (1955). Problems of structure and function in the amphibian oocyte nucleus. *Symposia Soc. Exptl. Biol.* 9, 358–370.

GALL, J. G. (1956). On the submicroscopic structure of chromosomes. *Brookhaven Symposia in Biol.* **8**, 17–32.

GALL, J. G. (1958). Chromosomal differentiation. *In* "The Chemical Basis of Development" (W. D. McElroy and B. Glass, eds.), pp. 103–135. Johns Hopkins Press, Baltimore, Maryland.

GALL, J. G. (1962). Unpublished observations.

GALL, J. G. (1963). The kinetics of DNase action on chromosomes. *Nature* in press.

GALL, J. G., AND CALLAN, H. G. (1962). H³ uridine incorporation in lampbrush chromosomes. *Proc. Natl. Acad. Sci. U.S.* **48**, 562–570.

HARRIS, H. (1959). Turnover of nuclear and cytoplasmic ribonucleic acid in two types of animal cell, with some further observations on the nucleolus. *Biochem. J.* **73**, 362–369.

HEITZ, E., AND BAUER, H. (1933). Beweise für die Chromosomenstruktur der Kernschleifen in den Knäuelkernen von *Bibio hortulanus* L (cytologische Untersuchungen an Dipteran, I). *Z. Zellforsch.* **17**, 67–82.

JACOB, F., AND MONOD, J. (1961). Genetic regulatory mechanisms in the synthesis of proteins. *J. Mol. Biol.* **3**, 318–356.

KROEGER, H. (1960). The induction of new puffing patterns by transplantation of salivary gland nuclei into egg cytoplasm of *Drosophilia. Chromosoma* **11**, 129–145.

MACGREGOR, H. C. (1963). Physiological variability in the oocytes of the crested newt. In preparation.

MACGREGOR, H. C., AND CALLAN, H. G. (1962). The actions of enzymes on lampbrush chromosomes. *Quart. J. Microscop. Sci.* **103**, 173–203.

McCLINTOCK, B. (1934). The relation of a particular chromosomal element to the development of the nucleoli in *Zea mays. Z. Zellforsch.* **21**, 294–328.

McMASTER-KAYE, R. (1962). The metabolism of nuclear ribonucleic acid in salivary glands of *Drosophilia repleta. J. Histochem. and Cytochem.* **10**, 154–161.

McMASTER-KAYE, R., AND TAYLOR, J. H. (1958). Evidence for two metabolically distinct types of ribonucleic acid in chromatin and nucleoli. *J. Biophys. Biochem. Cytol.* **4**, 5–11.

MECHELKE, F. (1953). Reversible Strukturmodifikationen der Speicheldrüsenchromosomen von *Acricotopus lucidus. Chromosoma* **5**, 511–543.

MILLER, O. L. (1962). Studies on the ultrastructure and metabolism of nucleoli in amphibian oocytes. *Proc. 5th Intern. Congr. on Electron Microscopy, Philadelphia, 1962* NN-8.

MIRSKY, A. E., AND RIS, H. (1951). The desoxyribonucleic acid content of animal cells and its evolutionary significance. *J. Gen. Physiol.* **34**, 451–462.

PRESCOTT, D. (1960). Nuclear function and nuclear-cytoplasmic interactions. *Ann. Rev. Physiol.* **22**, 17–44.

RUDKIN, G. T., AND CORLETTE, S. L. (1957). Disproportionate synthesis of DNA in a polytene chromosome region. *Proc. Natl. Acad. Sci. U.S.* **43**, 964–968.

SCHUMAKER, V. N., RICHARDS, E. G., AND SCHACHMAN, H. K. (1956). A study of the kinetics of the enzymatic digestion of deoxyribonucleic acid. *J. Am. Chem. Soc.* **78**, 4230–4236.

SIBATANI, A., DEKLOET, S. R., ALLFREY, V. G., AND MIRSKY, A. E. (1962). Isolation of a nuclear RNA fraction resembling DNA in its base composition. *Proc. Natl. Acad. Sci. U.S.* **48**, 471–477.

STICH, H., AND NAYLOR, J. (1958). Variation of desoxyribonucleic acid content of specific chromosome regions. *Exptl. Cell Research* **14**, 442–445.

SWIFT, H. (1958). Studies on nucleolar function. *In* "Symposium on Molecular Biology" (R. E. Zirkle, ed.), pp. 266–303. Univ. of Chicago Press, Chicago, Illinois.

SWIFT, H. (1962). Nucleic acids and cell morphology in dipteran salivary glands. *In* "The Molecular Control of Cellular Activity" (J. M. Allen, ed.), pp. 73–125. McGraw-Hill, New York.

TAYLOR, J. H., ed. (1963). Molecular models for organization of DNA into chromosomes. *In* "Molecular Genetics," Vol. I, Chapter II. Academic Press, New York.

THOMAS, C. A. (1956). The enzymatic degradation of desoxyribose nucleic acid. *J. Am. Chem. Soc.* **78**, 1861–1868.

VINCENT, W. S. (1955). Structure and chemistry of nucleoli. *Intern. Rev. Cytol.* **4**, 269–298.

The Plastids: Their Morphological and Chemical Differentiation

S. GRANICK

The Rockefeller Institute, New York, New York

I. Introduction

Studies on chloroplasts are of interest for two reasons. First, they are the cytoplasmic bodies that carry on photosynthesis. Second, they appear to be self-duplicating bodies which although under the control of the nucleus still possess some autonomy. Photosynthesis is the function of the mature chloroplast. How the mature chloroplast of higher plants arises, what factors govern its growth, multiplication, and differentiation will be our main concern. Some of the topics which I shall summarize and attempt to bring to date have been reviewed recently in a chapter in "The Cell" (Granick, 1961a) to which the reader is referred for a more complete account and for literature references. In addition I shall discuss two aspects of chloroplasts which have been studied recently in our laboratory, namely, steps in chlorophyll biosynthesis, and factors in the plastid inheritance of *Euglena*. Recent comprehensive reviews on chloroplast structure and function are those by Thomas (1961) and Menke (1962b).

II. Structure of the Mature Chloroplast

The chloroplast is only one of a number of plastid types found in the mature tissues of higher plants. There are yellow plastids of the carrot root (chromoplasts) which contain carotenoids in such high concentration (20–25% of their dry weight) that they are often crystalline. There are colorless amyloplastids which are specialized to store starch as in the potato tuber; proteinoplasts which store protein as in *Phajus grandiflorus* where the plastid contains a cluster of parallel needlelike protein crystals; and elaeoplasts which store oil. In addition, there are tiny colorless plastids such as those present in the epidermal cells of a leaf or in the root. One may ask: Do all of these types of plastids arise from one precursor and become specialized or does each arise from a different type of precursor? To answer this question we shall have to examine the development and details of the structure of the plastids.

First let us examine the structure of a mature chloroplast of a leaf. In the light microscope the chloroplast appears as a lens-shaped green body, about 5 μ long and 2–3 μ thick, which lies embedded in cytoplasm. The chloroplast may be homogeneous or contain fine granules, the grana. In the electron microscope a suspension of spinach chloroplasts mounted without treatment on a grid is seen to contain dense grana about 0.5–0.8 μ in diameter (Fig. 1). A granum is made up of a stack of 10–100 disks one on top of another like a stack of coins. When a granum is spread apart the individual disks may be shadowed with gold and readily made visible (Fig. 2). Finer details of the chloroplast are revealed after fixation with OsO₄, leaching with organic solvents, embedding in a polymer, and sectioning. What remains in the fine sections are mainly delicate membranous structures. These are summarized

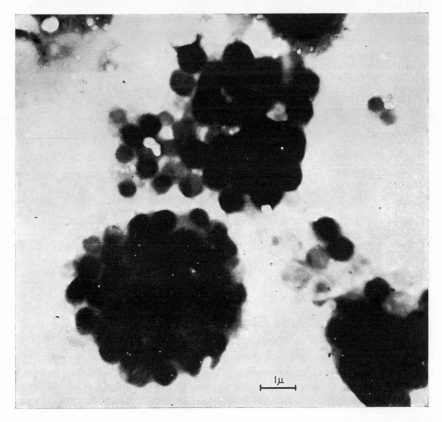

Fɪɢ. 1. Electron micrograph of isolated spinach chloroplast containing 40–60 dense bodies, the grana. (Granick and Porter, 1947.)

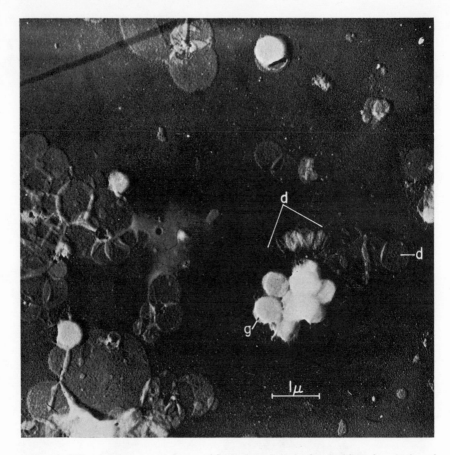

Fig. 2. Electron micrograph of grana (*g*) of a fragmented spinach chloroplast shadowed with gold. In the lower center is a dense cluster of grana. Above is a series of overlapping disks (*d*) and a pair of disks. (Granick and Porter, 1947.)

and interpreted in a diagram by von Wettstein (1959) (Fig. 3). Surrounding the chloroplast is a semipermeable double membrane, each 35–50 Å thick. Then there is a stroma material, consisting of a protein solution represented by protein molecules and fine granules. Embedded in the stroma are two types of membranes that are arranged in pairs. In the electron microscope a disk of a granum is seen to be made up of a pair of membranes or lamellae, each about 30 Å thick, comparatively dense to the electron beam, and enclosing a narrow space about 100 Å thick. The paired membranes that lie in the stroma and run from one granum to another are usually less dense. Some

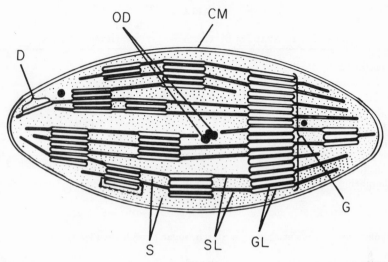

FIG. 3. Diagram of a barley chloroplast. *CM* is a cytoplasmic double membrane. *D* is a disk consisting of two membranes or lamellae. *OD* = osmiophylic droplet; *G* = granum or stack of disks; *S* = stroma region; *SL* = stroma lamella; *GL* = grana lamella. (von Wettstein, 1959.)

of the disks appear to be continuous throughout the plastid, i.e. extending throughout the stroma and grana regions. Disks in the grana region appear to be cemented to each other; plastids swollen in water show enlargement of the intradisk spaces (Eriksson *et al.*, 1961). Weier and Thomson (1962) have observed in some of their material from tobacco leaves that the membranes in the stroma region may not be continuous but fenestrated and may also appear as flattened ribbons or frets interconnecting the grana.

The approximate analysis of chloroplasts indicates that 35–55% or more of the dry weight is protein and 20–30% is lipid (Table I). In contrast to mitochondria the galactolipid content is high. About 9% of the dry weight of the chloroplast is chlorophyll a + b. Analyses of the double membranes of the chloroplasts which contain chlorophyll (Park and Pon, 1961) indicate that there is 1 chlorophyll molecule (mol. wt. 900) per $\sim$3000-mol-wt. protein (or per $\sim$25 amino acids). The analytical results depend on the method of chloroplast isolation. Recently Behren's technique has been applied to the isolation of chloroplasts from the frozen-dried state using organic solvents. The water-soluble components including enzymes are not washed away. With this method ribulose diphosphate carboxylase was demonstrated to be a chloroplast constituent (Smillie and Fuller, 1959); it was shown that bean and tobacco chloroplasts contained over 40% of the total leaf K^+,

TABLE I

Approximate Analysis of Chloroplasts of Higher Plants[a]

Constituent	% of dry weight[e]	Components		
Proteins	35–55	About 80% is insoluble		
Lipids	20–30		[b]Fats	50%
			Sterols	20
		[c]Choline 46%	Waxes	16
Galactolipids[c]	6	Inositol 22	Phosphatides	2–7
Phospholipids[c]	1.5	Glycerol 22		
		Ethanolamine 8		
		Serine 0.7		
Carbohydrate	Variable	Starch, sugar phosphates (3–7 C)		
Chlorophyll	∼9	Chlorophyll a 75%		
		Chlorophyll b 25%		
Carotenoids	∼4.5	Xanthophyll 75%		
		Carotene 25%		
Nucleic acids				
RNA	2–3			
DNA	0.5(?)			
Cytochromes				
f	∼0.1			
b_6	—			
Vitamins				
K	0.004			
E	0.08			
Plastoquinone	0.5			
Ash[d]	∼3			
Fe	0.1			
Cu	0.01			
Mn	0.016			
Zn	0.007			
P	0.3			

[a] From Granick (1961).
[b] On runner bean leaves (Eberhardt and Kates, 1957).
[c] On chloroplasts (Benson 1961, Wintermanns, 1960).
[d] On broken chloroplasts (Warburg, 1949).
[e] The water content of the chloroplast is 70-80% (Menke, 1962b).

Mg^{++}, and Ca^{++} and up to 70% of the total leaf N (Stocking and Ongun, 1962); and whole spinach chloroplasts were found to contain 7% of their dry weight as ribonucleic acid (RNA) and 0.5% as deoxyribonucleic acid (DNA) (Biggins and Park, 1962). Differential centrifugation in a sucrose-glycerol gradient has been applied to the separation of chloroplasts from mitochondria (Leech and Ellis, 1961); the absence of transaminase activity in mature chloroplasts was shown in this way.

In Table II are summarized some of the enzyme systems that are assumed or have been shown to occur in proplastids and chloroplasts. Thomas (1960) has compiled a list of the enzymes that have been found in chloroplasts.

III. Relation of Chloroplast Structure to Photosynthetic Function

In photosynthesis two main series of reactions occur, a light and a dark series. In the light, light quanta are absorbed by chlorophyll, the energy is transmitted probably by exciton migration through other chlorophylls of a photosynthetic unit (containing about 100–300 chlorophyll molecules) to an energy sink where water is split into unknown products, and O_2 is released. The unknown products XH (at the potential of the H_2 electrode?) and YOH

TABLE II

ENZYME SYSTEMS OF PLASTIDS[a]

A. Postulated enzyme systems in duplicating proplastid used for the synthesis of
   1. RNA; DNA
   2. Protein
   3. Lipid
   4. Starch

B. Postulated enzyme systems in differentiating plastid used for the synthesis of
   1. Protein: (a) structural, (b) enzymatic
   2. Lipid
   3. Chlorophylls
   4. Carotenoids

C. Known enzymes in mature plastids that function in photosynthesis
   1. Enzymes coupled with the photodecomposition of water, reduction of TPN and DPN, oxidation of cytochrome f, formation of O_2
   2. Enzymes that form ATP by photophosphorylation
   3. Enzymes that fix CO_2 in Ru-1,5diP by the Calvin cycle using TPNH or DPNH and ATP
   4. Enzymes of the pentose-P shunt that form Ru-5P indirectly from G-3P
   5. Enzymes that form starch indirectly from G-3P

[a] From Granick, 1961a.

(at the potential of the O_2 electrode) are in turn converted in a series of unknown steps into the form of adenosine triphosphate (ATP) and reduced triphosphopyridine nucleotide (TPNH), i.e. useful forms of chemical free energy. In the dark ATP and TPNH are used to convert (fix) CO_2 into starch, amino acids, fats, etc.

Fractionation of the steps in photosynthesis as related to structure began in 1937 when Hill (Hill and Wittingham, 1955) isolated chloroplasts and observed that O_2 could be produced in the light in the presence of an electron acceptor like ferric oxalate. In 1954 Arnon and co-workers showed that both the light and the dark reactions could occur in the isolated chloroplast. Trebst *et al.* (1958) found that when the soluble enzymes were leached out from the isolated chloroplast, the green residue could still photophosphorylate (i.e. form ATP from Pi and adenosine diphosphate (ADP) in the light), could still form O_2 in the light in the Hill reaction in the presence of suitable oxidants, and could still reduce TPN to TPNH on the addition of photosynthetic pyridine nucleotide reductase, but could not fix CO_2.

The colorless enzymes and the coenzymes that were leached out were important for CO_2 fixation. When these were added back, CO_2 fixation occurred. Through the work from Calvin's laboratory (Bassham and Calvin, 1961) it was found that CO_2 fixation required the substrate ribulose diphosphate, the enzyme ribulose diphosphate carboxylase, other 3–7-carbon phosphorylated sugars, the enzymes of the pentose phosphate shunt, and photosynthetic pyridine nucleotide reductase. [Photosynthetic pyridine nucleotide reductase = methemoglobin reductase = ferredoxin (Tagawa and Arnon, 1962; Appella and San Pietro, 1962; Mortenson *et al.*, 1962).] Other soluble components of the chloroplasts when added back to the leached chloroplasts also enhanced CO_2 fixation. These included substances that enhanced photophosphorylation like ascorbic acid, coenzyme K, coenzyme Q, Mg^{++}, K^+ in addition to other substances like pyridine nucleotides, ATP, and small amounts of phosphorylated sugars (Trebst *et al.*, 1958).

On the basis of these investigations CO_2 fixation is considered to occur in the stroma region which contains the soluble enzymes; in this region starch is formed from hexose phosphate produced via the Calvin cycle. The conversion of light into the chemical free energy of ATP and TPNH is considered to take place in the insoluble disk region of the chloroplast.

Where in the disk region does the light reaction occur? Thomas *et al.* (1953) had observed that insoluble chloroplast fragments of 100-Å diameter, which contained about 200 chlorophyll molecules, when illuminated with intense light could form O_2 in the Hill reaction at a rate 60% of that of whole chloroplasts. Park and Pon (1961) carried these observations further (Fig. 4).

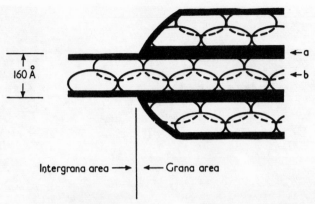

FIG. 4. Model for the lamellar structure within a spinach chloroplast. (a) Osmium-staining layer of the lamellar structure. Thickness 30 Å in the intergrana regions and 60 Å in the grana regions. (b) Particles forming the granular inner surface of the two layers making up the lamellar structure. The packing of oblate spheres would not be as simple as illustrated in the figure since the central axis of both layers would not be in the same vertical plane shown here. (Park and Pon, 1961.)

They sonically disintegrated the plastids and found that it was the double membrane structures (i.e. disks) which could carry out the Hill reaction in the light, and also CO_2 fixation if the soluble enzymes were added to these membranes. They observed granules 100×200 Å on the inside surfaces of the pair of double membranes.

A tempting hypothesis proposed by Park and Pon (1961, 1962) is to consider that these granules, "quantasomes," represent the elementary photosynthetic units where light energy is absorbed and converted to chemical energy. Assuming that the disks are essentially the granules closely packed in monolayers (Figs. 4, 5), the approximate composition of these granules may be estimated. A granule, 100×200-Å spheroid, is a lipoprotein, half lipid and half protein, of about 1 million molecular weight. Each granule contains $\sim$100 chlorophyll molecules and $\sim$25 carotenoid molecules. On the assumption of 1 manganese per granule, the granule contains 1 Mn atom (connected with O_2 evolution) and 6 Fe atoms (inclusive of 1 cytochrome f, 1 cytochrome b_6, and ferredoxin). In the granule is an energy sink which on the basis of electric dichroism (Sauer and Calvin, 1962) is suggested to consist of about 10 chlorophyll 690-mμ molecules and possibly the cytochromes also, all oriented with their porphyrin planes parallel to each other. Independent evidence for oriented chlorophyll 695 mμ has also been obtained by Olson *et al.* (1961, 1962); with the chloroplast of *Euglena* viewed on edge from above, the absorption of 695-mμ light was greatest and

FIG. 5. Granules of the disk membrane as represented in Fig. 4. On the right side, the upper membrane with its granules has been torn away. (By P. Healy; through the courtesy of M. Calvin).

the fluorescence emission was greatest when the vibration direction of the electric vector of the light was parallel to the long axis of the chloroplast and therefore parallel to the long edges of the chloroplast lamellae. The inference is made that the plane of the chlorophyll molecule lies parallel to the plane of the lamella. Some enzyme reactions connected with the energy sink to convert light energy to chemical energy might also be localized in the granule. An analogy might be made with the complex granule of similar molecular weight, α-ketoglutarate oxidase, which carries through about 5 enzyme reactions. It will be important to isolate the granules cleanly and study further their composition and enzymic contents.

In this connection the findings of Smith (1961) may be noted. Production of O_2 was found to begin in leaves when only a trace of chlorophyll ($\sim 1\%$ of the concentration found in a fully grown leaf as estimated by Goedheer, 1961) had been produced, at most not more than double the protochlorophyll initially present in the etiolated leaves. It will be of great interest to examine the granules of the disk at a time when they contain one or only a few chlorophyll molecules to note whether they can carry out the Hill reaction and to determine the absorption maximum of the chlorophyll at this time.

There are many questions left unanswered in such a simple quantasome model. For example, what is the function resulting from organization of the granules into membranes? Do the membranes of the disks function to separate oxidizable from reducible components? Do water-soluble materials like ATP and TPNH diffuse to the stroma regions *via* the intradisk spaces, and O_2 via the fatty interdisk regions? Are the double membranes required for phosphorylation as suggested by Green (1961) for the mitochondrial cristae? Do the carotenoids, besides protecting against the photo-oxidation of chlorophyll, function in electron transport, phosphorylation, etc.? Do the surfactant lipids function both as structural membrane components and in metabolism (Benson, 1961)?

IV. Morphological Changes in the Development of the Proplastid to the Chloroplast

In the higher plants, chloroplasts develop from tiny colorless or pale yellow bodies, the proplastids. This development may be divided into three overlapping phases (Fig. 6).

A. *The Phase of Proplastid Multiplication*

In rapidly dividing meristematic cells, at the stem tip the metabolic activity is predominantly that of DNA synthesis. Here 7–20 amoeboidlike proplastids (0.4–0.9 μ diameter) may be present per meristematic cell. The

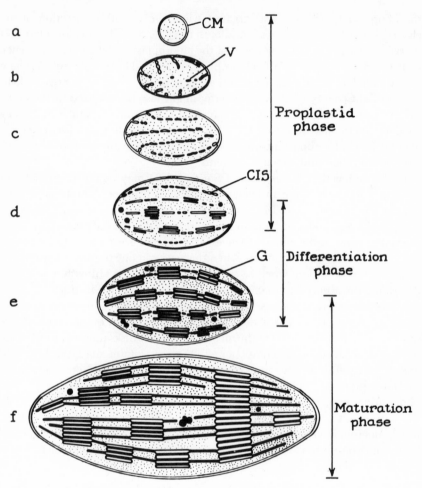

FIG. 6. Stages of development in the light. *CM* = double chloroplast membrane; *V* = vesicle; *CIS* = flattened cisternum; *G* = granum. (von Wettstein, 1959.)

proplastid is surrounded by two membranes, each $\sim$35 Å thick. Inside the proplastid is a granular background matrix or stroma. At an early stage spherical or elongate tubules (250–400 Å in diameter) grow out as invaginations of the inner membrane (Mühlethaler and Frey-Wyssling, 1959). When the cells are in the light, these tubules fuse and give rise, it is not yet clear how, to a pile of 4–15 flattened cisternae that lie parallel to each other but separate. The amoeboid proplastid may elongate and pinch off to form two proplastids. It is estimated that a proplastid may undergo a total of 3–4 such fissions. Starch grains, tiny oil droplets, and traces of carotenoids may be present. In *Lupinus* most of the meristematic cells of the leaf are formed along with their proplastids after the leaf emerges from the bud (Sunderland, 1960).

In this meristematic stage it may be difficult to distinguish tiny proplastids from mitochondria. However, in favorable cases, the youngest proplastids are usually larger, and contain one or a few flat cisternae, and starch (Menke, 1960). Starch synthesis has never been observed to occur in mitochondria. In algae where plastid growth is not repressed the distinction between chloroplasts and mitochondria is quite clear. A striking example is furnished by the recent electron microscope study of Manton (1959) on the tiny unicellar marine alga *Chromalina pulsilla*. This organism has a large chloroplast and a smaller mitochondrion. At cell division both the chloroplast and mitochondrion were seen to divide by fission.

B. *Phase of Differentiation*

In the elongating cells back of the meristematic stem tip, in the light the predominant syntheses are those of proteins, lipids, and RNA. In this phase the proplastids enlarge, become lens-shaped, and lose their amoeboid character. The flattened cisternae extend out in a plane and become disks. It is not certain whether all the membranes arise from the invaginations of plastid membranes or form *de novo* in the stroma and contribute to the growth of the cisternae, or whether a cisternum may invaginate at the edge to form two disks by longitudinal fission. Perhaps all of these processes occur. Certain regions of the cisternal membranes may increase in disk number faster than others to form the grana regions, or grana regions may develop before complete fusion of vesicles and cisternae has occurred. Concomitant with the development of the disks and grana regions is an increase in chlorophyll. Development of grana is not essential for chlorophyll formation. The bundle sheath plastids of corn contain disks that extend across the plastid but no grana regions, yet small amounts of chlorophyll are present in them (Hodge *et al.*, 1956)

C. *Phase of Maturation*

The chloroplast enlarges to final size, the grana become better defined, and fission ceases. By the time cells of tomato leaves have expanded to one-third their final length, the plastids have almost stopped dividing. During this time enzymes that will partake in photosynthesis also increase (Smillie and Fuller, 1960). Amino acid activating enzymes have been isolated from mature spinach chloroplasts (Bové and Raacke, 1959), presumably having served for protein synthesis of the plastids in their development.

V. The Effect of Light on Plastid Development

The differentiation of the proplastid to the chloroplast is controlled by factors of both the external and internal environment. The factors of the external environment, i.e. the cell culture environment, are light, temperature, water, and the components of the culture medium such as minerals, sugars, and diffusible products from adjacent and distant cells.

Light has a considerable effect on differentiation of proplastids apart from its role in photosynthesis. When seedlings are germinated in the absence of light, the proplastid of the etiolated leaf buds out vesicles from the inner proplastid membrane but the rate of fusion of the vesicles to form the flattened cisternae is slow. As a consequence the vesicles accumulate in a cluster to form a dense tiny body called the proplastid center or prolamellar body (Fig. 7). In barley seedlings grown in the dark for 3–10 days, the proplastids continue to enlarge and the cluster of vesicles appears to fuse to form a crystal-like lattice structure discovered independently by Heitz and by Leyon. According to Menke (1962a) this structure consists of tubules wound in spirals embedded in a matrix and ordered into a hexagonal array. The cross-section diameter of the tubule is 160 Å and the cross section of the spiral 450 Å. At the edges of the "crystal lattice" the tubules are connected to short or long vesicles extending out into the stroma.

When the etiolated barley leaf is exposed to 660- or 445-mμ light of very low intensity, sufficient to convert the protochlorophyllide-holochrome complex of the crystal lattice to chlorophyllide (see below), the crystal lattice appears to fall apart into tiny vesicles (Eriksson *et al.*, 1961). This process can occur at 3° C. If, while still at 3° C. the leaves are exposed to light of higher intensity, the vesicles disperse and may organize into double membranes that appear as flattened cisternae or primary layers, or they may appear as a series of spheres one inside another (i.e. resembling layers of onion scales) (Klein, 1960). These geometric rearrangements induced by increased light appear to be essentially physical. Further changes require enzyme action, for they take place only at a higher temperature, e.g. 20° C; and they require ATP, for they are inhibited by dinitrophenol. Overlapping the protochlorophyllide-to-chlorophyllide effect is probably a red-infrared light effect (Hendricks, 1960), which once set off by the conversion of phytochrome 650 mμ to phytochrome 730 mμ may continue in the dark for several hours or days. It seems to cause a decrease in cytoplasmic viscosity or an increase in cytoplasmic streaming. The red-infrared effect may also be involved in the induction of rapid protein synthesis. Finally, light of increased intensity, perhaps used for photosynthesis, may be required to bring about an increase in the number of disks and the fusion of the disks with each other to form grana.

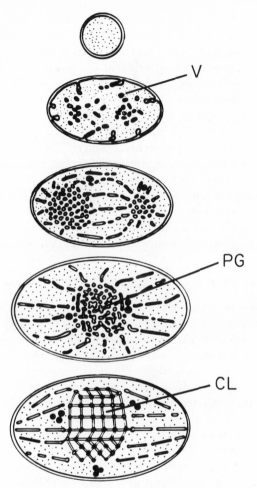

Fig. 7. Stages of development in the dark. V = vesicle; PG = proplastid center or prolamellar body; CL = crystal lattice. (von Wettstein, 1958.)

Chlorophyll synthesis occurs rapidly at this time. In the barley mutant *Xantha*, no chlorophyll is formed and plastid development stops at the stage of the primary layer formation before the disks organize into grana (von Wettstein, 1958).

Although the growth of the etiolated barley leaf and of the proplastids may occur in the dark, the growth of etiolated bean leaves and its proplastids in the dark is quite limited. Bean seedlings grown in darkness have very small pale yellow leaves and are responsive to a red-infrared system. When red light of 650 mμ is used to illuminate the bean leaves, 2 minutes daily for 4 days, the leaves expand (Liverman *et al.*, 1955), the plastids double in volume, the plastid N doubles and reaches the level of the mature chloroplast, the lipid content reaches two-thirds that of the mature chloroplast, and a sixfold increase in carotenoids occurs. No grana or chlorophyll develops under these conditions. Light of 730 mμ inhibits the effect of the 650-mμ light (Mego and Jagendorf, 1961). When bean proplastids containing prolamellar bodies are isolated from the cells and exposed to $\frac{1}{2}$ minute of white light, the tubules of the lattice seem to elongate (Klein and Poljakoff-Maybeer, 1961). The reorganization of materials under the influence of the red-infrared effect is reminiscent of this light effect on protoplasmic streaming and of chloroplast orientation in certain algae (Haupt and Thiele, 1961).

It is of interest to note that in enlarging barley root cells back of the meristem, the proplastid may contain one or several vesicles or even a stunted crystal lattice structure. Exposure to light does not cause these proplastids to green.

VI. Chlorophyll Biosynthesis in the Developing Plastid

Our recent studies on chlorophyll biosynthesis in barley furnish additional information on plastid differentiation (Granick, 1960, 1961b). For some time now we have been investigating the biosynthetic chain of chlorophyll. From studies on *Chlorella* mutants it was found that both heme and chlorophyll are made along the same biosynthetic chain to protoporphyrin. Then iron is inserted into protoporphyrin to make heme, the prosthetic group of the cytochromes, and Mg is inserted into protoporphyrin to form Mg protoporphyrin, which is converted in a series of steps to chlorophyll (Fig. 8). The mitochondria contain the cytochromes a, c_1, and b. The chloroplasts contain the cytochromes b_6, and f.

With the light and fluorescence microscope we have observed that the etiolated barley proplastid begins as a smooth tiny spheroid (Fig. 7). As it enlarges a dense yellowish granule appears in the center. This granule fluoresces very faintly pink when excited with ultraviolet light (Strugger and Kriger, 1960). When δ-aminolevulinic acid (a porphyrin precursor) is fed to these leaves in the dark, the intensity of red fluorescence in the granule is greatly enhanced. This granule is the prolamellar body or crystal lattice structure. Its yellow color is due to carotenoid and the red fluorescence is due to protochlorophyllide (Mg VP). Whether these pigments are in or on the tubules of the crystal lattice structure is not known. The fact that carotenes and porphyrins are found to be localized in the crystal lattice suggests that at least the last steps in their synthesis if not all of them occur in the proplastid.

Protoporphyrin (PROTO) $\xrightarrow{\text{?}}$ $\begin{cases} \text{Phycoerythrin} \\ \text{Phycocyanin} \end{cases}$

↓

Mg protoporphyrin (Mg PROTO)

↓

Mg protoporphyrin monomethyl ester

┆ 4 steps ?

↓

Mg vinyl pheoporphyrin-a$_5$ (Mg VP)
(protochlorophyllide) $\left[630 \ m\mu \right]$

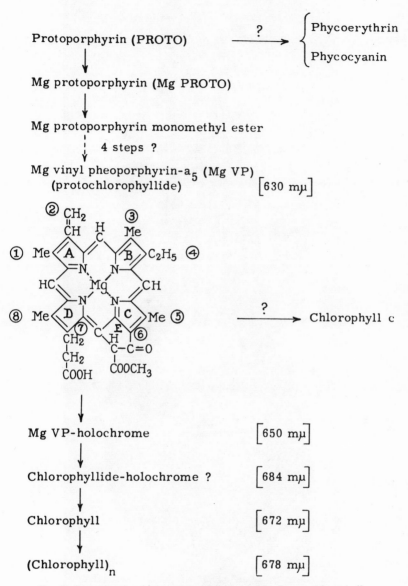

$\xrightarrow{\text{?}}$ Chlorophyll c

↓

Mg VP-holochrome $\left[650 \ m\mu \right]$

↓

Chlorophyllide-holochrome ? $\left[684 \ m\mu \right]$

↓

Chlorophyll $\left[672 \ m\mu \right]$

↓

(Chlorophyll)$_n$ $\left[678 \ m\mu \right]$

FIG. 8. Steps in the biosynthesis of protoporphyrin to chlorophyll.

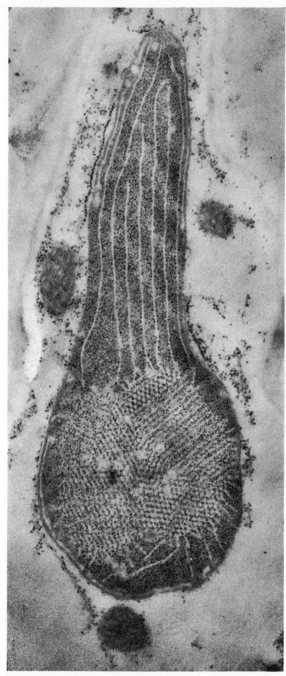

FIG. 9. Etiolated corn proplastid fixed in formalin, stained with uranyl acetate. 43,000×. (A. Jacobson, unpublished.) Ribosomes are present in the crystal lattice region (lower half) and also in the lamellar region.

This means that the enzymes that synthesize these compounds may all be present in the early proplastid.

When etiolated barley leaves are first exposed to red light, there is a lag of several hours before chlorophyll synthesis begins which then continues at a steady rate (Smith and Coomber, 1959; Virgin, 1958). If, however, the etiolated leaves are first exposed to red light of 650 mμ for a few minutes, then placed in the dark for several hours, then exposed to continuous white or red light, chlorophyll will be formed without a time lag. One may postulate the following changes that occur after the brief illumination and during the subsequent dark period of several hours. The spiral tubules in the crystal lattice unwind and fragment into vesicles which organize into cisternae. At the same time protein synthesis appears to have been stimulated so that structural as well as enzymic proteins are formed. On continuous illumination these changes then permit chlorophyll to be synthesized at a steady rate. More disks form, the membranes become more osmiophilic and compacted into grana regions.

As mentioned before, 650-mμ light will convert protochlorophyllide in the crystal lattice or vesicles into chlorophyllide. It only requires low intensity light to do this because the quantum yield for this process is about 0.5–1 and the reduction can occur even at $-60°$ C (Smith, 1960). The reduction consists in the introduction of two H atoms *trans* to each other at the double bond of the D ring. Smith has estimated the molecular weight of the holochrome to be about one million. On feeding δ-aminolevulinic acid to etiolated bean leaves, we have been able to detect three forms of protochlorophyllide (Mg VP) by means of absorption spectrum changes. A large amount of Mg VP is formed, indeed so large that the yellow leaves become greenish, not due to chlorophyll, but to Mg VP. One form of Mg VP, the principal one, has an absorption maximum at 631 mμ; it is readily bleached by light and is not converted to chlorophyllide. Another form has an absorption maximum at 650 mμ (the holochrome of Smith) and is converted to chlorophyllide. The third form has a maximum at 650 mμ but is not converted by light; it is estimated to represent as much as half of the total 650-mμ pigment. This third form may be a protein-Mg VP holochrome that lacks a reducing constituent required for photoreduction of the Mg VP.

Etiolated barley leaves fed δ-aminolevulinic acid in the dark form about 10 times their normal content of Mg VP. The Mg VP is localized in the proplastid. When these greenish leaves are exposed to light of low intensity for 24 hours, insufficient to result in photosensitized bleaching, the Mg VP content is decreased only slightly, whereas chlorophyll is formed abundantly yet not more than in controls that had not been treated with δ-aminolevulinic

acid. The excessive Mg VP (631 mμ) in the plastids is scarcely used in chlorophyll formation; nor does it inhibit chlorophyll formation, i.e. Mg VP (63 1mμ)
does not serve as feedback control. Because excessive Mg VP is not converted
to chlorophyllide in the light, and the content of Mg VP (650 mμ) is not
appreciably increased, either the holochrome protein of the 650-mμ form is
limiting or its reducing constituent is limiting. Perhaps the photochemical
reduction of protochlorophyllide to chlorophyllide controls the synthesis of
δ-aminolevulinic acid. The enzymatic property of the holochrome has not
yet been demonstrated.

After chlorophyllide has been formed in the leaf by brief exposure to light,
the enzymic esterification with phytol to form chlorophyll may proceed in
the dark to be completed within 10–60 minutes (Smith, 1960; Wolf and
Price, 1957; Godnev and Akulovich, 1961). During this time progressive
changes in the absorption maxima are observed. The following tentative
interpretations may be made of these changes. Chlorophyllide when formed
and while still attached to the holochrome has an absorption maximum at
685 mμ or somewhat higher (Smith, 1960; Shibata, 1957). When the chlorophyllide separates from the holochrome its absorption maximum becomes
673 mμ. After about 10 minutes at 20° chlorophyllide 673 mμ becomes esteri-
fied with phytol to chlorophyll 673 mμ which is in the monomolecular form.
These changes may occur while the membranes are still cisternae. Concomitant with esterification the chlorophyll may become organized into the disk
membrane (or the intradisk granule?) where carotenoid is present, so that
light energy may now be transferred to a small extent from carotenoid to
chlorophyll (Butler, 1960). The chlorophyllide 685-mμ and 673-mμ forms are
both very labile to photobleaching (Goedheer, 1961). The 673-mμ form
usually becomes photostable within 15 minutes, possibly indicative of its becoming phytylated to chlorophyll and then associated with and protected by
the carotenoids (Stanier, 1960). On the basis of high chlorophyll fluorescence
at first, and a gradual decrease in fluorescence, Goedheer suggested that new
chlorophyll molecules did not add to form into units of high pigment density,
but rather that as more chlorophyll molecules were formed the distances
between them gradually decreased during greening. In the grana of the
mature leaf the absorption maximum is at 678 mμ and the fluorescence is
decreased, suggestive of an interaction with neighboring chlorophyll molecules. The 678-mμ maximum may represent an overlapping of 673-mμ chlorophylls with the oriented 695-mμ chlorophylls of the quantasomes. Goedheer
has shown that chlorophyll 673 mμ is capable of photosynthesis provided
that very high intensities of light are used. Smith (1954, 1961) considered
that no more chlorophyll than that equivalent to twice the original amount

of protochlorophyllide in an etiolated leaf was needed for O_2 to be detected in photosynthesis. Similar shifts in the maxima of chlorophyll, as followed by fluorescence changes during greening, have been observed by Krasnovskii (1961). Kok (1961) reported a chlorophyll with an absorption maximum at 700 mμ in a concentration 1/300 that of normal chlorophyll; he suggested that this 700-mμ pigment may represent a chlorophyll at the energy sink. It would be interesting to look for such a pigment in etiolated leaves that have formed a small amount of 673-mμ chlorophyll and which can form O_2 in photosynthesis.

VII. Inheritable Factors That Control the Plastids

The internal factors that control plastids may be divided into three groups. (1) There are the inheritable factors that reside in the nucleus, the genes, which segregate according to Mendelian laws. (2) There are the inheritable factors that reside in the plastids. (3) There are other inheritable factors which reside in the cytoplasm exclusive of the plastids. About 20 cases of cytoplasmic inheritance have been reviewed by Caspari (1948).

We shall not discuss the effect of nuclear genes on the phenotypic expression of the plastids, except to note that many genes are known to affect chlorophyll formation or plastid development. How these gene effects are brought about, whether by defects in syntheses of those specific enzymes formed in the cytoplasm that will be incorporated into the plastid, or by control of key synthetic reactions of the plastid with inhibitors, is not known.

Evidence for inherited cytoplasmic factors outside of plastids has been obtained by interspecies crosses of *Epilobium* (Michaelis, 1954) and from studies on variegated plants such as *Mirabilis jalapa* var. *albomaculata* (Correns, 1909).

Evidence for inherited plastid factors is as follows:

1. Plastids arise only from pre-existing plastids or proplastids. This result is best seen in algal cells which have one chloroplast that divides by fission prior to cell division. It is well documented both in vegetative and in meiotic divisions.

2. A plastid carries its own inheritable individuality. This fact is best demonstrated by finding a cell with one chloroplast that is recognizably different from the others, although under the influence of the same nucleus and cytoplasm. In *Spirogyra triformis*, van Wisselingh (1920) observed that in a cell in addition to normal chloroplasts which contained pyrenoids there was a chloroplast which lacked normal pyrenoids. In subsequent cell divisions this abnormal chloroplast continued to persist in the presence of the normal ones. If a mutation had occurred in the nucleus or cytoplasm that affected pyrenoids, it should have affected all plastids alike.

3. In crosses in some variegated higher plants usually only the plastids of the female cytoplasm are inherited, that is, maternal inheritance occurs. By making reciprocal crosses, using flowers growing on a branch that has color-less plastids and flowers growing on a branch that has green plastids, the inheritance of green or white plastids may be observed. The green plastids were found to be different from the white plastids even though the internal environment of cytoplasm and nucleus in the cells in which they developed was identical. Such a study is represented by the classic work of Bauer (1930) on *Pelargonium zonale* var. *albomarginata*.

4. Plastids from closely related species may be shown to be different even though they develop in cells with the presumably identical cytoplasm and nucleus. In interspecies crosses of *Oenothera* (Schwemmle, 1938) two kinds of green plastids from the two species responded differently in the same cyto-plasm and nucleus, i.e. one type of plastid grew green and the other yellow or white. This means that chloroplasts from one species are different from chloroplasts of another species in terms of their inheritable qualities. It is reasonable to consider that during evolution, species arose which not only had modified their nuclear genes but also their inheritable plastid factors.

5. Evidence for the presence of RNA and possibly also of DNA in the plas-tids suggests that these molecules might represent the heritable factors or templates of plastids. For example, the chloroplasts of Turkish tobacco, washed with trichloroacetic acid and extracted with alcohol-ether were found to contain per unit dry weight 3% RNA, 0.7% DNA, and 9–11% N (Cooper and Loring, 1957). Spinach chloroplasts isolated by the nonaqueous method contained 7% of their dry weight as RNA and 0.5% as DNA (Biggins and Park, 1962). In Fig. 9, the proplastid of corn is seen to be filled with large numbers of granules which are presumably the ribosomes since they can be digested away by ribonuclease. In *Acetabularia*, after the removal of the nucleus, chloroplast multiplication and protein synthesis continued for 21–28 days; Naora *et al.* (1960) considered that the synthesis of chloroplast RNA occurred during this time.

Other studies suggest that DNA also is in the plastids. Ris and Plaut (1962) found several Feulgen-positive bodies in the chloroplast of *Chlamydomonas* which became negative after deoxyribonuclease (DNase) treatment. They also observed 25–30-Å thick fibrils in these regions which were absent after DNase treatment. In *Euglena*, Sagan and Scher (1961) fed tritiated thymidine; the labeling was found to be in the cytoplasm, presumably in the plastids, and was removed by DNase. Likewise, Stocking and Gifford (1959) observed labeling of *Spirogyra* plastids after feeding with tritiated thymidine.

The evidence that plastids come from pre-existing ones, and the more recent evidence for the presence of RNA and DNA in these bodies, suggest that plastids may be more autonomous than has been suspected.

VIII. Cytoplasmic Inheritance of *Euglena* Plastids

One of the most often quoted examples of cytoplasmic inheritance is the effect of streptomycin on *Euglena*. When *Euglena gracilis* var. *bacillaris* was treated for even a few hours with a dose of streptomycin, 50–100 times less than would kill the cell, and the streptomycin was washed away, it was found that the progeny of these cells would no longer grow green in the light; they were permanently bleached (Provasoli *et al.*, 1948). Other means of bleaching these cells irreversibly were found to be a mild dose of ultraviolet light, or growth at 35° C for a few cell divisions. This bleaching was interpreted as "curing" the cells of chloroplasts. I should like to discuss this example of cytoplasmic inheritance somewhat fully not only because a number of laboratories including our own have begun to study it systematically, but also because it demonstrates the complexities of the problems involved.

Recent studies by Gibor and Granick (1962a) have indicated that the inheritable change observed is not due to a loss of plastids, but rather due to an inability of the proplastids to differentiate into chloroplasts. Evidence that the bleached cells contain proplastids has been obtained from studies of normal green *Euglena* cells grown in the dark, from spontaneous mutants, and from streptomycin- or ultraviolet-induced mutants.

A normal green *Euglena* cell when grown under continuous light on a rich medium forms a chloroplast chain of fused chloroplasts from which it is difficult to determine the number of chloroplasts per cell. When grown in the dark on a rich medium *Euglena* loses its chlorophyll. Now if after about 2 weeks of growth in the dark the cell is given about 5 hours of light to develop a little chlorophyll and examined in the fluorescence and phase microscope, the number of small discrete fluorescent proplastids can be counted. In some cells 10 proplastids are found which, prior to cell division, pull apart so that each daughter cell contains 10 proplastids. The division and segregation are too precise to be considered random and would repay investigation.

This *Euglena* strain was found by Robbins *et al.* (1953) to give rise to spontaneous "mutants" at a rate of 0.1% and in our hands with a different medium at a rate of 1–2%. The series of spontaneous mutants which was isolated ranged all the way from white cells, to cells in which the plastids contained traces of carotenoids, to pale green cells, and also to cells that grew at a slower rate so that petite-white and petite-green colonies were formed.

When *Euglena* was treated with streptomycin the colonies that arose were all "bleached." Among the "bleached" colonies, intermediates were recognized that ranged from white through pale yellow and differed in their ability to form porphyrins from δ-aminolevulinic acid. Gibor has demonstrated the presence of proplastids in these cells and has also observed intermediates in the population of ultraviolet-bleached cells (Table III). The criteria that were used for the identification of proplastids were (1) the presence of carotenoids in these 2-μ bodies; (2) the accumulation of porphyrin in them when the cells were incubated with δ-aminolevulinic acid; (3) the association of proplastids with paramylum; (4) the morphologic similarity of proplastids with the proplastids of normal dark-grown cells.

Recent studies, by Epstein and Schiff (1960), Lyman *et al.* (1961) and Schiff *et al.* (1961), of Brandeis University, on ultraviolet bleaching of *Euglena* have revealed a number of intriguing findings. The nonlethal dose of ultraviolet which produced bleached cells had an action spectrum with a peak at 260 and a shoulder at 280 mμ. They interpreted this to signify that the ultraviolet-sensitive material was a nucleoprotein. On the basis of hit theory they calculated that there might be about 30 ultraviolet-sensitive units per cell, all of which would need to be damaged before the cell would bleach. They found also that the ultraviolet damaging effect could be reversed by illumination of the cells with light in the range of 350–450 mμ; the photoreactivation at moderate light intensity depended on the total dose, not dose rate. This property to become photoreactivatable would remain for a week if the ultraviolet-irradiated cells were kept in the dark and prevented from

TABLE III

SYNTHETIC ABILITIES OF PLASTIDS OF EUGLENA STRAINS[a]

Plastids	Strain	Carotenoids	Porphyrin fluorescence		Chlorophyll
			No ALA[b]	Added ALA	
I	U. V.	+	White	White	−
	P. W. 3	+	White	White	−
II	Streptomycin	++	White	Red	−
	35° C	++	White	Red	−
III	B6Ln7W	+++	Red	Red	− (Occasional)
IV	Normal	+++	Red	Red	+

[a] From Gibor and Granick, 1962.
[b] ALA = δ-aminolevulinic acid.

multiplying, but the property was lost if the cells went through 5 or 6 divisions. If the ultraviolet-irradiated cells were illuminated with red light of 650 mμ they were not photoreactivated; these cells became green in red light perhaps because there was a store of enzymes in the cells sufficient for one generation. Such cells gave rise only to bleached colonies. The Brandeis group considered that ultraviolet-sensitive units responsible for chloroplast formation were damaged, so could not be replicated, and became diluted out among the progeny. Because green light-grown cells were only half as sensitive to ultraviolet bleaching, they suggested that the ultraviolet-sensitive units were in the cytoplasm rather than in the nucleus, and most probably in the plastids themselves.

One of the simplest hypotheses to explain these results is to assume that ultraviolet light acts upon especially sensitive nucleoprotein particles (of this *Euglena* strain), which function to replicate and also to make one or more of the enzymes required for the differentiation of the proplastid to chloroplast. To explain Gibor's findings of intermediate types of bleached cells, the ultraviolet-sensitive unit may have to be considered as not a single but a multifactor unit, i.e. containing several kinds of genes. Such an assumption is being tested by the Brandeis group with quantitative ultraviolet and photoreactivation studies.

Gibor and Granick (1962b) have recently demonstrated that the ultraviolet-sensitive units are in the cytoplasm and not in the nucleus. By the use of an ultraviolet microbeam apparatus which could irradiate only the nucleus, or only the cytoplasm but not the nucleus, it was found that bleached cells were obtained if the cytoplasm was irradiated with ultraviolet light. On the other hand, if the nucleus was irradiated with small doses, or doses sufficiently large to prevent cell multiplication, no bleached cells arose. Because the bleached cells are not cured in subsequent cell divisions, it is inferred that the U.V.-sensitive units do not arise in the nucleus.

There is some evidence for both RNA and DNA in *Euglena* plastids. Pogo and co-workers (1962) found that the RNA of green cells had a slightly higher proportion of adenylic and uridylic and a slightly lower proportion of cytidylic and guanylic acids than etiolated or bleached cells. Sagan and Scher (1961) on the basis of autoradiography, with *Euglena* treated with tritiated thymidine and subsequently with nucleases, concluded that some DNA was present in *Euglena* chloroplasts.

The general properties of plastids may permit us to understand and predict the properties of mitochondria (Rouiller, 1960) and vice versa. In connection with inherited factors a few interesting findings on mitochondria may be mentioned. The mitochondrion of *Trypanosoma mega* has been shown by M.

Steinert to arise in connection with the kinetoplast, which at an early stage consists of a Feulgen-positive moiety and a mitochondrial portion that later enlarges and separates from the Feulgen-positive body (quoted by Mirsky and Osawa, 1961). In animal tissue culture, decreasing the temperature to 16° for a day revealed, by the Feulgen and tritiated thymidine techniques, that DNA was present in mitochondria (Chevremont *et al.*, 1961). Interest in small amounts of DNA that may be present in the cytoplasm and that may have been overlooked before has been stimulated by the finding of DNA polymerase in the cytoplasm (Prescott *et al.*, 1962). The crystalline cytochrome b_2 of baker's yeast (in mitochondria?) contains 1 heme, 1 riboflavin, and a 15 deoxyribose polynucleotide chain; its DNA portion is not essential for lactic dehydrogenase activity; its $(A + T)/(G + C)$ ratio is 2.6, but the ratio for yeast DNA is 1.79 (Morton *et al.*, 1961). It has not been established that cytochrome b_2 is in the yeast mitochondria. In yeast, when given a short pulse of radioactive phosphate, it is found that the portion of RNA that turns over rapidly is complementary to total DNA of the yeast in terms of the nucleotide ratios, but the bulk of the yeast RNA, presumably inclusive of the cytoplasm and mitochondria, is different (Yčas and Vincent, 1960). Evidence for a half-life of only 10 days has been obtained for mitochondria of liver nuclei (Fletcher and Sanadi, 1961). A *de novo* origin of mitochondria and proplastids as blebs from the nuclear membrane has been proposed in studies on the egg cell in the archegonium of *Pteridium aquilinum* (Mühlethaler and Bell, 1962).

IX. Origin of Plastid Types

In the beginning of this paper the different types of plastid that were encountered in higher plants were summarized. The phenotypic expression and functioning of a differentiated cell is in part due to its content of differentiated cytoplasmic bodies, e.g. in this case its content of a particular type of plastid. To explain the origin of a differentiated plant cell it is necessary to explain the origin of a particular type of plastid from a proplastid. We have seen that the proplastid contains starch, protein, carotenoids, protochlorophyllide, and probably many of the enzymes for the synthesis of these compounds, together with RNA and possibly DNA. The simplest hypothesis to explain the plastid types is to assume that they all arise from one kind of proplastid and pass through successive phases of development, which can be arrested at a particular phase. Along this pathway one or several biosynthetic chains may be inhibited at the expense of one that may be released from control. A scheme based on this hypothesis is presented in Fig. 10. In seedling roots, the proplastids in the zone of elongation become leucoplasts which do not develop

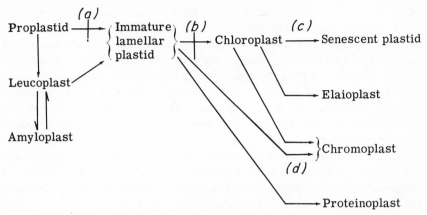

FIG. 10. Relations of plastid types. (a) Arrested in normal root cells and in leaves of some barley and corn mutants. (b) Arrested in normal leaf epidermal cells. (c) Degeneration in autumn leaves; hypertrophy of some *Oenothera* plastids. (d) Chromoplasts of buttercup petals.

further even in the presence of light; they may contain a granular protein matrix, lipid droplets, and tiny starch grains. Amyloplasts are characteristic of the starch storage organs such as the potato tuber. A mature chloroplast may develop from such an amyloplast. In epidermal cells plastid development is arrested at the immature vesicle or cisternal stage. Chromoplasts may form from mature chloroplasts or from immature pale green plastids containing a few grana. The chromoplasts appear to represent a form of fatty plastid degeneration in which fats and carotenoids increase and finally break down to release yellow oily droplets into the cytoplasm (Frey-Wyssling and Kreutzer, 1958).

It is conceivable that during evolution, for efficiency, functions like photosynthesis and respiration with their own complexity of enzymes were segregated into bodies, the plastids and the mitochondria, along with the inheritable units required for their duplication and differentiation. Nuclear genes could still maintain a control on the duplication and differentiation of these bodies by way of specific inhibitors (Yoshida, 1962). Tests of this hypothesis could be made more readily if plastids could be maintained in *in vitro* culture.

X. Summary

The structure of the mature chloroplast has been reviewed briefly and an attempt has been made to relate its structure to its functions in photosynthesis. The hypothesis of the quantasome is a recent exciting development and one will await with interest the analyses of the isolated granules of the disks to know more of their organization and enzymic content. It may be that, like the α-ketoglutarate dehydrogenase complex of the mitochondria which contains five enzymes, this quantasome unit will have all the enzymes one seeks that aid in the conversion of light to chemical free energy.

The morphological and chlorophyll changes as the proplastid differentiates into the chloroplast have been outlined. These changes are a fascinating gear-meshing of many chemical synthetic reactions, at least three of which are controlled or triggered by light.

Finally, the evidence has been discussed for the partial autonomy of the plastid in the cell, and as an example the interesting system in *Euglena* that controls proplastid differentiation has been considered. In addition to this nucleoprotein ultraviolet-sensitive system there must be an inheritable system that governs proplastid multiplication itself, of which we know very little.

The material presented, though diffuse, already suggests the importance of knowing more about cell organelles like the plastid and mitochondrion if we are to understand cell differentiation.

References

APPELLA, E., AND SAN PIETRO, A. (1962). Physical properties of photosynthetic pyridine nucleotide reductase. *Biochem. Biophys. Research Communs.* 6, 349.

ARNON, D. I., WHATLEY, F. R., AND ALLEN, M. B. (1954). Photosynthesis by isolated chloroplasts. II. Photosynthetic phosphorylation, the conversion of light into phosphate bond energy. *J. Am. Chem. Soc.* 76, 6324.

BASSHAM, J. A., AND CALVIN, M. (1961). The way of CO_2 in plant photosynthesis. *Proc. 5th Intern. Congr. Biochem., Moscow,* Symposium No. 6, Preprint No. 48.

BAUER, E. (1930). "Einführung in die Vererbungslehre," 2nd ed., 431 pp. Bornträger, Berlin.

BENSON, A. A. (1961). Lipid function in photosynthetic structures. *In* "Light and Life" (W. D. McElroy and B. Glass, eds.), p. 392. Johns Hopkins Press, Baltimore.

BIGGINS, J., AND PARK, R. B. (1962). Nucleic acid content of the chloroplasts of *Spinacae oleracea.* Bio-Organic Chem. Quart. Rept. U CRL 9900, p. 39.

BOVÉ, J., AND RAACKE, J. D. (1959). Amino acid activity enzymes in isolated chloroplasts from spinach leaves. *Arch. Biochem. Biophys.* 85, 521.

BUTLER, W. L. (1960). Energy transfer in developing chloroplasts. *Biophys. Biochem. Research Communs.* 2, 419.

CASPARI, E. (1948). Cytoplasmic inheritance. *Advances in Genet.* 2, 1.

CHEVREMONT, M., BASSLEER, R., AND BAECKELAND, E. (1961). Nouvelles recherches sur les acides désoxyribonucléiques dans les cultures de fibroblastes refroidées puis réchauffées. Etudes cytophotométrique et histoautoradiographique localisation cytoplasmique d'ADN. *Arch. biol.* 72, 501.

COOPER, W. D., AND LORING, H. S. (1957). The ribonucleic acid composition and phosphorus distribution of chloroplasts from normal and diseased Turkish tobacco plants. *J. Biol. Chem.* **228**, 813.

CORRENS, C. (1909). Vererbungsversuche mit blass (gelb) grünen und bunt blättrigen Sippen bei Mirabilis jalapa, Urtica pilulifera und Lunaria annua. *Z. induktive Abstammungs- u. Verebungslehre* **1**, 291.

EBERHARDT, F. M., AND KATES, M. (1957). Incorporation of C^{14} or P^{32} into the phosphatides of runner bean leaves. *Can. J. Bot.* **35**, 907.

EPSTEIN, H. T., AND SCHIFF, J. A. (1961). Studies of chloroplast development in *Euglena*. 4. Electron and fluorescence microscopy of the protoplastid and its development into a mature chloroplast. *J. Protozool.* **8**, 427.

ERIKSSON, G., KAHN, A., WALLES, B., AND VON WETTSTEIN, D. (1961). Zur makromolekularen Physiologie der Chloroplasten. III. *Ber. deut. botan. Ges.* **74**, 221.

FLETCHER, M. J., AND SANADI, D. R. (1961). Turnover of rat liver mitochondria. *Biochim. et Biophys. Acta* **51**, 356.

FREY-WYSSLING, A., AND KREUTZER, E. (1958). The submicroscopic development of chloroplasts in the fruit of *Capsicum annum* L. *J. Ultrastruct. Research* **1**, 397.

GIBOR, A., AND GRANICK, S. (1962a). The plastid system of normal and bleached *Euglena gracilis J. Protozool.* **9**, 327.

GIBOR, A., AND GRANICK, S. (1962b). Ultraviolet sensitive factors in the cytoplasm that affect the differentiation of *Euglena* plastids. *J. Cell Biol.* **15**, 599.

GODNEV, T. N., AND AKULOVICH, N. K. (1960). Pigments of the protochlorophyll group. *Doklady Akad. Nauk SSSR* **134**, 710. (English transl.)

GOEDHEER, J. C. (1961). Effect of changes in chlorophyll concentration on photosynthetic properties. I. Fluorescence and absorption of greening bean leaves. *Biochim. et Biophys. Acta* **51**, 494.

GRANICK, S. (1960). Magnesium porphyrins in chlorophyll biosynthesis. *Federation Proc.* **19**, 330.

GRANICK, S. (1961a). The chloroplasts: inheritance, structure and formation. *In* "The Cell" (J. Brachet and A. E. Mirsky, eds.), Vol. II, p. 948. Academic Press, New York.

GRANICK, S. (1961b). The pigments of the biosynthetic chain of chlorophyll and their interactions with light. *Proc. 5th Intern. Congr. Biochem., Moscow*, Symposium No. 6, Preprint No. 65.

GRANICK, S., AND PORTER, K. R. (1947). The structure of the spinach chloroplast as interpreted with the electron microscope. *Am. J. Botany* **34**, 545.

GREEN, D. E. (1961). Structure and function of subcellular particles. *Proc. 5th Intern. Congr. Biochem., Moscow*, Plenary Lecture, Preprint No. 176.

HAUPT, W., AND THIELE, R. (1961). Chloroplastenbewegung bei Mesotaenium. *Planta* **56**, 388.

HENDRICKS, S. B. (1960). Photoreactions controlling photoperiodism and related responses. *In* "Comparative Biochemistry of Photoreactive Systems" (M. B. Allen, ed.), Academic Press, New York.

HILL, R., AND WITTINGHAM, C. D. (1955). "Photosynthesis." Wiley, New York.

HODGE, A. J., McLEAN, J. D., AND MERCER, F. V. (1956). A possible mechanism for the morphogenesis of lamellar systems in plant cells. *J. Biochem. Biophys. Cytol.* **2**, 597.

KLEIN, S. (1960). The effect of low temperature on the development of the lamellar system in chloroplasts. *J. Biochem. Biophys. Cytol.* **8**, 529.

KLEIN, S., AND POLJAKOFF-MAYBER, A. (1961). Fine structure and pigment conversion in isolated etiolated proplastids. *J. Biochem. Biophys. Cytol.* **11**, 433.

Kok. B. (1961). Significance of P_{700} as an intermediate in photosynthesis. *Proc. 5th Intern. Congr. Biochem., Moscow,* Symposium No. 6, Preprint No. 208.

Krasnovskii, A. A. (1961). Photochemistry of chlorophyll, the state and transformations of pigments of photosynthetic organisms. *Proc. 5th Intern. Congr. Biochem., Moscow,* Symposium No. 6, Preprint No. 25.

Leech, R. M., and Ellis, R. J. (1961). Coprecipitation of mitochondria and chloroplasts. *Nature* 190, 790.

Liverman, J. L., Johnson, M. P., and Starr, L. (1955). Reversible photoreaction controlling expansion of etiolated bean leaf disks. *Science* 121, 440.

Lyman, H., Epstein, H. T., and Schiff, J. A. (1961). Studies of chloroplast development in *Euglena* I. Inactivation of green colony formation by U.V. light. *Biochim. et Biophys. Acta* 50, 301.

Manton, I. (1959). Electron microscopical observations on a very small flagellate: the problem of *Chromulina pusilla* Butcher. *J. Marine Biol. Assoc. U.K.* 38, 319.

Mego, J. L., and Jagendorf, A. T. (1961). Effect of light on growth of black valentine bean plastids. *Biochim. et Biophys. Acta* 53, 237.

Menke, W. (1960). Einige Beobachtungen zur Entwicklungsgeschichte der Plastiden von Elodea canadensis. *Z. Naturforsch.* 15b, 800.

Menke, W. (1962a). Über die Struktur der Heitz-Leyonschen Kristalle. *Z. Naturforsch.* 17b, 188.

Menke, W. (1962b). Structure and chemistry of plastids. *Ann. Rev. Plant Physiol.* 13, 27.

Michaelis, P. (1954). Cytoplasmic inheritance in Epilobium and its theoretical significance. *Advances in Genet.* 5, 287.

Mirsky, A. E., and Osawa, J. (1961). The interphase nucleus. *In* "The Cell" (J. Brachet and A. E. Mirsky, eds.), Vol. II, p. 677. Academic Press, New York.

Mortenson, L. E., Valentine, C. R., and Carnahan, J. E. (1962). An electron transport factor from *Clostridium posteurianum. Biochem. Biophys. Research Communs.* 7, 448.

Morton, R. K., Armstrong, J. McD., and Appleby, C. A. (1961). The chemical and enzymic properties of cytochrome b_2 of bakers yeast. "Haematin Enzymes" (J. E. Falk, R. Lemberg, and R. K. Morton, eds.), p. 501. Pergamon Press, New York.

Mühlethaler, K., and Bell, P. R. (1962). Untersuchungen über die Kontinuität von Plastiden und Mitochondrien in der Eizelle von Pteridum aquilinum (L.) Kuhn. *Naturwiss.* 49, 63.

Mühlethaler, K., and Frey-Wyssling, A. (1959). Entwicklung und Struktur der Proplastiden. *J. Biophys. Biochem. Cytol.* 6, 507.

Murakami, S. (1962). *J. Cell Biol.* in press.

Naora, Hiroto, Naora, Hatsuko, and Brachet, J. (1960). Studies on independent synthesis of cytoplasm. Ribonucleic acids in *Acetabularia mediterranea. J. Gen. Physiol.* 43, 1083.

Olson, R. A., Butler, W. L., and Jennings, W. H. (1961). The orientation of chlorophyll molecules *in vivo*: Evidence from polarized fluorescence. *Biochim. et Biophys. Acta* 54, 615.

Olson, R. A., Butler, W. L., and Jennings, W. H. (1962). The orientation of chlorophyll molecules *in vivo. Biochim. et Biophys. Acta* 58, 144.

Park, R. B., and Pon, N. G. (1961). Correlation of structure with function in *Spinacea oleracea* chloroplasts. *J. Mol. Biol.* 3, 1.

PARK, R. B., AND PON, N. G. (1962). *J. Mol. Biol.* 4, in press.

POGO, A. O., BRAWERMAN, G., AND CHARGAFF, E. (1962). New ribonucleic acid species associated with the formation of the photosynthetic apparatus in *Euglena gracilis*. *Biochemistry* 1, 128.

PRESCOTT, D. M., BALLUM, F. J., AND KLUSS, B. C. (1962). Is DNA polymerase a cytoplasmic enzyme? *J. Cell Biol.* 13, 172.

PROVASOLI, L., HUTNER, S., AND SHATZ, A. (1948). Streptomycin induced chlorophyllless races of *Euglena*. *Proc. Soc. Exptl. Biol. Med.* 69, 279.

RIS, H., AND PLAUT, W. (1962). Ultrastructure of DNA-containing areas in the chloroplast of *Chlamydomonas*. *J. Cell Biol.* 13, 383.

ROBBINS, W. J., HERVEY, A., AND STEBBINS, M. E. (1953). *Euglena* and Vitamin B_{12}. *Ann. N.Y. Acad. Sci.* 56, 818.

ROULLIER, C. (1960). Physiological and pathological changes in mitochondrial morphology. *Intern. Rev. Cytol.* 9, 227.

SAGAN, L., AND SCHER, S. (1961). Evidence for cytoplasmic DNA in *Euglena gracilis*. *J. Protozool.* 8, Suppl., Abstract 20.

SAUER, K., AND CALVIN, M. (1962). Molecular orientation in quantasomes. *J. Mol. Biol.* 4, 451.

SCHIFF, J. A., LYMAN, H., AND EPSTEIN, H. T. (1961a). Studies of chloroplast development in *Euglena*. II. Photoreversal of the U.V. inhibition of green coloring formation. *Biochim. et Biophys. Acta* 50, 310.

SCHIFF, J. A., LYMAN, H., AND EPSTEIN, H. T. (1961b). Studies of chloroplast development in *Euglena*. III. Experimental separation of chloroplast development and chloroplast replication. *Biochim. et Biophys. Acta* 51, 340.

SCHWEMMLE, J. (1938). Genetische und zytologische Untersuchungen an Eu-Oenatheren, Teil I bis VI. *Z. Induktive Abstammungs- u. Verbungslehre* 75, 358.

SHIBATA, K. (1957). Spectroscopic studies on chlorophyll formation in intact leaves. *J. Biochem. (Tokyo)* 44, 147.

SMILLIE, R. M., AND FULLER, R. C. (1959). Ribulose 1,5-diphosphate carboxylase activity in relation to photosynthesis by intact leaves and isolated chloroplasts. *Plant Physiol.* 34, 651.

SMILLIE, R. M., AND FULLER, R. C. (1960). Photosynthetic and respiratory enzymes in leaves. *Federation Proc.* 19, 212.

SMITH, J. H. C. (1954). The development of chlorophyll and oxygen-evolving power in etiolated barley leaves when illuminated. *Plant. Physiol.* 29, 143.

SMITH, J. H. C. (1960). Protochlorophyll transformations. *In* "Comparative Biochemistry of Photoreactive Systems" (M. B. Allen, ed.), pp. 25–275. Academic Press, New York.

SMITH, J. H. C. (1961). *Proc. 5th Intern. Congr. Biochem., Moscow*, Symposium No. 6, Preprint No. 110.

SMITH, J. H. C., AND COOMBER, J. (1959). Absorption spectrum of protochlorophyll holochrome. *Carnegie Inst. Wash. Year Book* 58, 331.

STANIER, R. (1960). Carotenoid pigments: Problems of synthesis and function. *Harvey Lectures* 54, 219.

STOCKING, C. R., AND GIFFORD, E. M., JR. (1959). Incorporation of thymidine in chloroplasts of Spirogyra. *Biochem. Biophys. Research Communs.* 1, 159.

STOCKING, C. R., AND ONGUN, A. (1962). The intracellular distribution of some metallic elements in leaves. *Am. J. Botany* 49, 284.

STRUGGER, F., AND KRIGER, L. (1960). Untersuchungen über die Struktur der Plastiden etiolierten Pflanzen. *Protoplasma* 52, 230.

SUNDERLAND, N. (1960). Cell division and expansion in the growth of the leaf. *J. Exptl. Botany* 11, 68.

TAGAWA, K., AND ARNON, D. I. (1962). Ferredoxins as electron carriers in photosynthesis and in the biological production and consumption of hydrogen gas. *Nature* 195, 537.

THOMAS, J. B. (1960). Chloroplast structure. *In* "Encyclopedia of Plant Physiology" (W. Ruhland, ed.), Vol. 1, p. 511. Springer, Berlin.

THOMAS, J. B., BLAAUW, O. H., AND DUYSENS, L. N. M. (1953). On the relation between size and photochemical activity of fragments of spinach grana. *Biochim. et Biophys. Acta* 10, 230.

TREBST, A. V., TSUJIMOTO, H. Y., AND ARNON, D. I. (1958). Separation of light and dark phases in the photosynthesis of isolated chloroplasts. *Nature* 182, 351.

VIRGIN, H. I. (1958). Studies on the formation of protochlorophyll and chlorophyll a under varying light treatments. *Physiol. Plantarum* 11, 347.

WARBURG, O. (1949). "Heavy Metal Prosthetic Groups and Enzyme Action" (A. Lawson, transl.) Oxford Univ. Press., London and New York.

VAN WISSELINGH, C. (1920). Über Variabilität und Erblichkeit. *Z. Induktive Abstammungs- u. Vererbungslehre* 22, 65.

VON WETTSTEIN, D. (1958). The formation of plastid structures. *Broookhaven Symposia in Biol.* 11, 138.

VON WETTSTEIN, D. (1959). The effect of genetic factors on the submicroscopic structures of the chloroplast. *J. Ultrastruct. Research* 3, 234.

WEIER, T. E., AND THOMSON, W. W. (1962). The grana of starch-free chloroplasts of *Nicotiana rustica. J. Cell Biol.* 13, 89.

WINTERMANNS, J. F. G. M. (1960). Concentration of phosphatides and glycolipids in leaves and chloroplasts. *Biochim. Biophys. Acta* 44, 49.

WOLF, J. B., AND PRICE, L. (1957). Terminal steps of chlorophyll A biosynthesis in higher plants. *Arch. Biochem. Biophys.* 72, 293.

YČAS, M., AND VINCENT, W. S. (1960). A ribonucleic acid fraction from yeast related in composition to desoxyribonucleic acid. *Proc. Natl. Acad. Sci. U.S.* 46, 804.

YOSHIDA, Y. (1962). The nuclear control of chloroplast activity in *Elodea* leaf cells. *Protoplasma* 54, 476.

Organization and Disorganization of Extracellular Substances: The Collagen System

JEROME GROSS, CHARLES M. LAPIERE, AND
MARVIN L. TANZER

The Robert W. Lovett Memorial Group for the Study of Diseases Causing Deformities; Department of Medicine, Harvard Medical School and the Massachusetts General Hospital, Boston, Massachusetts

During growth and development, from embryo to senescence, the organism undergoes remodeling, radical and rapid early in life, more subtle and slow with increasing age. Reproducibility of form and the repetitive pattern of morphologic changes demand great precision of timing coupled with localization of synthesis, orientation of tissue constituents, and ordered removal of structural elements. Our understanding of the biosynthesis of macromolecules and of gene action in determining primary protein structure grows rapidly, although information is limited largely to soluble proteins. The processes of synthesis, orientation, and organization of structural elements such as fibrous proteins and lipoprotein lamellae are just beginning to attract concerted attention. Removal of such structural elements is of equal importance since organized growth is obviously not a simple matter of accretion. Dismantling of tissue components must be exquisitely controlled, both temporally and spatially, to permit growth and change without impeding function. Failure of this regulation very likely leads to malformation or altered physiology.

Our long-range goal is to gain understanding of the mechanisms and regulatory processes involved in organization, removal, and replacement of the structural elements of tissues during embryonic development, growth, and senescence. The strategy is to focus on a single well-characterized tissue element known to be involved in remodeling processes throughout life, using it as a thread to lead us through the labyrinth of relevant and irrelevant biochemical processes to the mechanisms controlling structural organization.

Even assuming a solution to the problem of macromolecular biosynthesis, many obvious questions lie on the next level of organization. For example, how do subunits of myosin and actin aggregate in the proper place, at the proper time, and in the proper amounts? A similar question may be asked of the lipid, polysaccharide, and protein constituents of the lamellar structures of cells. What are the mechanisms responsible for assembling and disassembling the mitotic apparatus? How is the deposition of extracellular fibers controlled so that the basic structural outline of a bone may be reproduced identically through generations, permitting extensive repetitive remodeling throughout the life cycle?

Polymerization of macromolecules in biological systems may often require enzymatic catalysis as in the fibrinogen-fibrin transformation. In other cases highly specific molecular interactions and internal changes in conformation may occur which require little energy transfer. They will happen because of the specific molecular structure which permits the formation or breaking of strategically placed secondary bonds (hydrogen, hydrophobic, electrostatic, etc.). These reactions will occur because they lead to more probable, hence stable, structures under a particular set of environmental conditions. Some of these reactions will be reversible, others not. Recent studies (such as that of Tompkins and Yielding, 1961) on enzyme systems indicate that association-dissociation reactions between the subunits of enzyme molecules may influence the type and extent of catalytic activity. In this particular instance glutamic dehydrogenase may be reversibly dissociated into smaller fragments *in vitro*, the dissociated subunits displaying alanine dehydrogenase activity. There is a growing list of enzymes which may be reversibly dissociated into subunits with changes in activity. It is possible that state of aggregation as well as intramolecular conformation changes will prove to be important regulatory mechanisms. The reversible *in vitro* precipitation of structural proteins from solution (collagen, fibrinogen, paramyosin, and actomyosin), in highly ordered fibrous arrays, and the formation of familiar looking membranes from solutions of lipids and proteins are appealing as indicators of basic mechanisms of *in vivo* organization. They also provide model systems for studying the forces involved in molecular aggregation.

Organization of Collagen

The extracellular protein collagen has proved to be a particularly interesting model for studies on morphogenesis at the molecular level. The behavior of this protein in solution probably represents the earliest demonstration of reversible, structurally specific aggregation *in vitro* (Gillette, 1872; Nageotte, 1927; Schmitt *et al.*, 1942; Gross *et al.*, 1954). Collagen is widely distributed throughout the animal kingdom and within the vertebrate organism where, as the structural framework of most tissues, it may account for as much as 30% of the total protein. It is involved in nearly all remodeling processes. Collagen is readily isolated, purified, and characterized, both in the solid state and in solution. Under physiologic conditions this protein is remarkably resistant to the known proteolytic enzymes. An experimentally useful proportion of the native fibrils may be dissolved, then reconstituted as fibers in various forms. Detailed discussions of collagen structure, chemistry, and biology may be found in a number of recent books and reviews (Bear, 1952; Randall, 1953; Gustavson, 1956; Tunbridge, 1956; Stainsby, 1958; Harrington and Von Hippel, 1961; Harkness, 1961).

The molecule itself is a long, relatively rigid rod about 300 Å in length and 15 Å wide, composed of 3 polypeptide chains, each in the configuration of a polyglycine or polyproline II helix (Rich and Crick, 1961). Figure 1 illustrates in a general way our picture of its molecular organization and mode of aggregation into fibrillar arrays. It is interesting to compare, in the figure, the dimensions of the collagen molecule (tropocollagen) with the dimensions of a 70-S ribosome drawn to scale. It is clear that this small subcellular organelle would have to go through some remarkable acrobatics to give birth to the monster in one piece! The picture is further complicated by the fact that one of the three polypeptide chains differs in amino acid composition from the other two (Piez et al., 1961). One may suppose that this huge molecule is built in separate subunits which, because of their highly specific structural organization, associate in the correct manner, either on a surface or in the intracellular fluid. Perhaps specific aggregates of ribosomes such as those recently implicated in hemoglobin synthesis (Warner et al., 1962) may also be involved here.

It is reasonable to assume that a complete collagen molecule (tropocollagen) is synthesized within the cell. The various subcellular fractions involved in protein synthesis have been isolated from collagen-synthesizing tissues such as carregeenan granuloma (Green and Lowther, 1959; Lowther et al., 1961) and the chick embryo (Prockop et al., 1961); tropocollagen capable of forming cross-striated fibrils in vitro has been obtained from the microsomal fraction by Green and Lowther (1959). Thi microsomal collagen, most highly labeled with C^{14}-proline, probably represents newly synthesized protein. Thus far, in vitro cell-free synthesis of collagen has not been achieved (Green and Lowther, 1959; C. Mitoma, personal communication). Recently Peterofsky and Udenfriend (1962) have reported incorporation of C^{14}-proline into hydroxyproline in a preparation of fortified chick embryo microsomes. It remains to be demonstrated that this is truly collagen-related hydroxyproline. It has been proposed (Stetten, 1944; Gould and

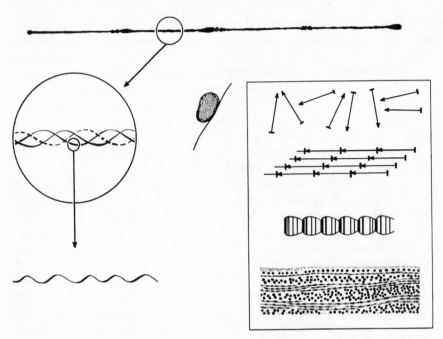

Fig. 1. Collagen fine structure. The tropocollagen molecule shown in the upper portion has an axial ratio of about 200:1, with an asymmetric fine structure along its length, roughly subdividing the molecule in quarters. It is compared with a 70-S ribosome on a membrane sketched to scale below it. The enlargement of the section of the molecule to the left illustrates the three polypeptide chains, one of which, the dashed line, is different in amino acid composition from the other two. A further enlargement of a section of one of the chains illustrates the helical configuration of each of the polypeptides. In the box at the right is a representation of the manner in which tropocollagen units are postulated to aggregate, overlapping each other in a staggered array by about one-quarter of their length, thereby giving rise to a collagen fibril with a repeating period of about 700 Å. The fine structure within each of these periods, illustrated in the fibril, would be a reflection of the fine details of asymmetry of the tropocollagen units arrayed in register. The lower portion of this block illustrates the manner in which collagen fibrils are found organized in plywood-like sheets in a variety of tissues.

Woessner, 1957; Hausmann and Neuman, 1961; Piez and Likins, 1957; Sinex *et al.*, 1959) that synthesis of the two unusual amino acids, hydroxyproline and hydroxylysine, is not completed until after their precursors, proline and lysine, are built into the peptide chain. However, there still exists the alternative possibility, that these two amino acids are hydroxylated in an activated form such as s-RNA proline or lysine and the product incorporated in the proper position in the collagen sequence as suggested by Robertson and associates (1959). Manner and Gould (1962) recently found a small amount of s-RNA hydroxyproline in chick embryo tissues. Although the search has been intensive, no clear-cut collagen intermediate, such as a single polypeptide chain or an incompletely hydroxylated peptide, has yet been found.

While it has been postulated (Schmitt, 1960) that some enzymatic process is required to "activate" the newly synthesized tropocollagen unit prior to fibril formation, similar to the fibrinogen-fibrin transformation, there is no evidence to suggest that catalysis is essential.

The manner in which collagen moves from the cell to intercellular space is not clear. Porter and Pappas (1959) have proposed the formation of thin "unit fibrils" near the cell surface which are extruded and grow by accretion of secreted collagen molecules. The release of intracellular collagen by shredding of the peripheral cytoplasm is suggested by electron microscope studies (Yardley et al., 1960; Karrer, 1960; Chapman, 1961). Stearns, years previously (1940), had reported fibril formation occurring about bits of cytoplasm left behind by wandering fibroblasts.

The course of maturation of collagen in the extracellular space has been followed by isotopic labeling and differential extraction with cold salt solutions of varying ionic strength and pH (Harkness et al., 1954; Jackson, 1957; Jackson and Bentley, 1960). The most recently formed protein, as indicated by the rate and extent of incorporation of labeled amino acids, is soluble in cold salt solutions at physiologic ionic strength and pH. Those fractions which require more concentrated salt solutions for extraction and, finally, low pH media, are progressively older. Figure 2 schematically illustrates this idea. The large bulk of collagen in an adult animal, in tissues which are not undergoing remodeling, may last for a lifetime (Neuberger and Slack, 1953; Thompson and Ballou, 1956). Truly old collagen is essentially insoluble and may be removed from the tissue only after denaturation.

This phenomenon of changing solubility with time may be duplicated in a systematic way *in vitro*. Collagen dissolved in cold neutral salt solutions may be polymerized to typical cross-striated fibrils by warming to body temperature (Gross et al., 1955; Jackson and Fessler, 1955; Gross and Kirk, 1958). Fibrils, so reconstituted, are readily solubilized on cooling but become progressively insoluble with the passage of time; within 2 weeks at body temperature such reconstituted fibrils are completely insoluble even at low pH (Gross, 1958, 1959, 1961b). This phenomenon might be explained by the effect of Brownian movement, permitting the relatively rigid and structurally specific molecules to fit together in the most stable association. The close association of side chains would enhance the formation of large numbers of

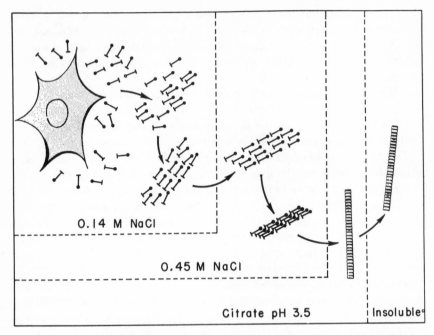

FIG. 2. Representation of a hypothesis explaining the significance of the different extractable collagen fractions. The rodlike units represent collagen molecules (tropocollagen). Cold physiological saline (0.14 M NaCl) extracts the most recently formed collagen (and perhaps also that resulting from physiological degradation); hypertonic salt solution extracts the same material plus older collagen in a more ordered state of aggregation; acid citrate buffer extracts all of the above plus some of the older collagen in the typical fibrillar form. The insoluble fibrils are older and their degree of cross-linking has prevented solubilization.

secondary-type bonds. Covalent links might also form with the passage of time. It is not inconceivable that a similar phenomenon might take place in other macromolecular systems in which aggregation is likely to occur. If we may speculate that certain enzyme systems are regulated by periodic or cyclic association-dissociation reactions, an increasing time of association might result in irreversible "sticking." Unless there is a rapid removal of such molecules, or some system of spacers to keep them apart, they might well be lost to the cell by reason of irreversible aggregation. The rate of loss would depend upon the rate of turnover relative to the duration of aggregation; the more rapidly molecules are removed and replaced, the shorter the association time and the smaller the opportunity for irreversible complexing. Possibly some of the irreversible features of differentiation involve such a process of "aging," particularly if nonreplaceable templates were involved.

Recent studies on the denaturation of the collagen molecule have brought to light the observation that intramolecular structure also changes with time, in terms of cross-linking between the three polypeptide chains. Orekhovitch and Shpikiter (1955) and Doty and Nishihara (1958) reported that gentle denaturation of acid-extracted collagen disrupted the molecule into two subunits as observed in the ultracentrifuge, one unit having about twice the molecular weight of the other. These components were found to be random coils as shown by viscosity and optical rotatory measurements. It was assumed at that time that both neutral-salt- and acid-extracted collagen had the same molecular organization. However, it has been recently observed (Orekhovitch *et al.*, 1959, Piez *et al.*, 1961) that neutral-extracted collagen, upon denaturation, gives rise to only one boundary in the ultracentrifuge, with a sedimentation coefficient similar to that of the lighter component of acid-extracted collagen.

The neutral-extracted collagen molecule can be dissociated into three separate polypeptide chains of about equal molecular weight, chromatographic fractionation revealing two major peaks and amino acid analysis indicating that one of the three chains, α_2, differs from the other two, α_1, in composition (Piez *et al.*, 1961) (Fig. 3). Acid-extracted collagen, on the other hand, contains dimers in addition to the two types of α monomers. One dimer, β_2, consists of a pair of α_1 chains. The other, β_1, is composed of one α_1 and one α_2 chain. A third and much smaller fraction of the denatured collagen seems to consist of trimers, in which all three chains are cross-linked (called γ by Altgelt *et al.* (1961)) (Fig. 4). Isotope incorporation studies (Orekhovitch *et al.*, 1960; Piez *et al.*, 1961) show that the two α chains are labeled rapidly and equally, the specific activity being manyfold greater than in the β components even 24 hours after administration of the isotope. With passage of time β_1 and β_2 become more heavily labeled, suggesting that formation of dimers is a process of "maturation," probably occurring outside the cell. The cross links holding the chain pairs together have the strength of covalent bonds since they are not broken by high concentrations of hot urea or guanidine.

It has been possible to demonstrate the *in vitro* production of such cross links by simply incubating lyophilized neutral-salt-extracted collagen at 37° in a moist atmosphere for several days (Gross, Piez, and Martin unpublished). The amount of dimerization occurring *in vitro* under these conditions has not exceeded 30% (estimated by ultracentrifuge patterns) of that found in native acid-extracted collagen. The bonds that are formed under these conditions

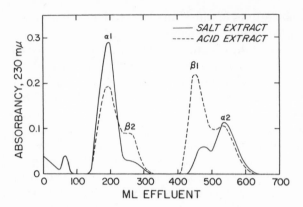

FIG. 3. Elution patterns of denatured rat skin collagen, chromatographed on CM-cellulose at 40° C. Solid line, 20 mg salt-extracted collagen; dashed line, 20 mg acid-extracted collagen. A linear gradient between 0.07 and 0.17-μ acetate buffer (pH 4.83) was employed. The labels refer to both patterns (Piez *et al.*, 1962).

are as stable as those of the native acid-extracted collagen, suggesting that they are similar. It may be that such cross-links would form spontaneously if reactive groups on adjacent polypeptide chains were held in close approximation for relatively long periods of time at body temperature. Transesterification may play a role in this process since ester-like bonds are thought to exist in collagen (Gallop *et al.*, 1959; Blumenfeld and Gallop, 1962). There is even the possibility that such reactions might be accelerated by adjacent groups which might have catalytic activity.

The *in vitro* aggregation behavior of collagen molecules has provided some interesting examples of interactions, influenced by a variety of nonspecific agents. These agents have the pronounced effect of producing several different collagen aggregates with highly specific organization (Fig. 5). Some years ago (Highberger *et al.*, 1951; Schmitt *et al.*, 1953; Gross *et al.*, 1954; Gross, 1956) it was shown that in the presence of serum acid glycoprotein, collagen fibrils reconstituted with a period of 3000 Å in contrast to the native period of 640 Å. With adenosine triphosphate (ATP) in the solution, collagen precipitated in the form of short crystallites, also 3000 Å in length, but with an entirely different intraperiod fine structure. Different salt concentrations could induce fibrils to form with periods of 220 Å, 640 Å, or a completely amorphous structure. It became evident that the specificity for producing these highly organized aggregates resided in the collagen molecules themselves and not in the inducing factors (Gross *et al.*, 1952). A single macromolecular species is capable of organizing in a number of grossly different,

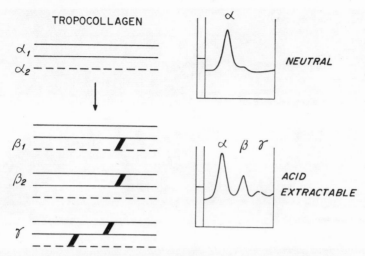

Fig. 4. Intramolecular cross-linking of collagen. At the top is newly formed molecule (neutral-salt-extracted) with no cross links between the three polypeptide chains. To the right is the ultracentrifuge diagram of this fraction, denatured by heating. The three polypeptide chains are completely separated, giving rise to a single boundary since they are all about the same molecular weight. A small amount of contaminating dimer is shown by the low, faster moving peak. With the passage of time dimers are formed by cross-linking between the chains. Two different dimers, β_1, and β_2, and the trimer γ are found in the acid-extracted collagen, in addition to the individual chains. The sedimentation pattern resulting from denaturation of the older acid-extracted collagen is shown on the right of the lower portion of the diagram.

but highly ordered, forms, and the differences in organization depend primarily upon the manner in which the molecules associated.

By a short flight of imagination one might suppose that other macromolecular systems may be capable of a variety of aggregation patterns, which, if associated with enzymatic or other activity, might be capable of a broad functional range. Variation in relatively "nonspecific" factors such as ionic strength, pH, and concentration of other low molecular weight substances might be sufficient to accomplish macromolecular rearrangements of considerable functional significance.

Aside from the biologic implications, studies on these particular modes of aggregation led to fairly accurate predictions of the size and shape of the collagen molecule (Gross *et al.*, 1954) and have allowed a more detailed structural analysis of the "charge profile" and other details of molecular structure (Schmitt, 1959; Hodge and Schmitt, 1960; Hodge, 1960). Another interesting aspect of the biologic significance of the structural specificity of

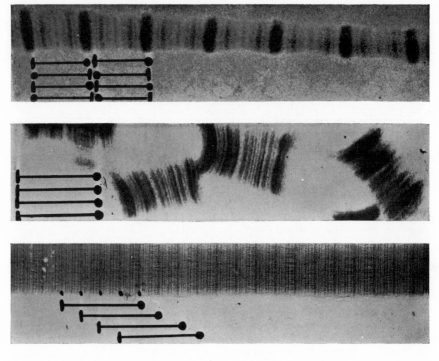

Fig. 5. Electron micrographs of three forms in which a solution of collagen may be reconstituted. *Top:* Fibrous long spacing (FLS) with an axial period of about 3000 Å. Molecules oriented at random to give a symmetric intraperiod fine structure. *Center:* Segment long spacing (SLS). Molecules oriented parallel and all facing in the same direction with no overlap. *Bottom:* Native type fibril with 640-Å period. Molecules all facing in same direction but overlapping by about one-quarter their length.

collagen is the observation of Glimcher *et al.* (1957), indicating that the formation of hydroxyapatite crystals will occur on collagen fibrils *in vitro* only if these fibrils are reconstituted in the native form, i.e., with the 640-Å periodicity.

One of the more baffling features of collagen organization is the mechanism whereby fibrils are arranged in plywood-like lattices, as found in the entire dermis of marine animals, including the whale (Mizuhira, 1951). This type of framework is also found in specialized organs of land animals, for example, the cornea (Jakus, 1956, 1960). Elsewhere, as in skin and bone, discrete local regions are characterized by the same type of orthogonally oriented lattice work of fibrils.

In the skin of amphibian larvae the pattern of organization is as highly ordered as in fish skin. During metamorphosis, however, there is a rapid reorganization and proliferation of the larval dermis to resemble the more random interlacing form characteristic of land animals (Kemp, 1961). This remarkable geometric array of fibrils in relatively acellular areas has attracted the attention of a number of investigators (Garrault, 1937; Niizima *et al.*, 1954; Weiss and Ferris, 1956; Weiss, 1957; Gross, 1961a). Edds and Sweeny (1961) have discussed this problem in some detail at a previous Growth Symposium. None of the postulated mechanisms for this extraordinary reorganization carries the weight of experimental evidence.

Experimental Lathyrism

In attempting to approach the problem of remodeling, we have examined a chemically induced malformation of skeletal structure. The experimental disease lathyrism is produced in a wide variety of growing animals by feeding the common sweet pea *Lathyrus odoratus* (Geiger *et al.*, 1933) or by giving β-aminopropionitrile (βAPN), the active compound from this plant (see Selye, 1957, for review). The major morphologic alterations produced by βAPN are severe skeletal deformities (Fig. 6), abdominal hernias, and dissection of the aorta (Ponseti and Shepard, 1954). The application of βAPN, aminoacetonitrile, and semicarbazide to the chick embryo results in dosage-dependent and time-dependent progressive increase in tissue fragility, quantitatively paralleled by a large rise in collagen extractable in cold neutral salt solutions (Levene and Gross, 1959). Figure 7 illustrates the linear rise in the extractable collagen obtained from long bones during the first 12 hours after administration of βAPN (Tanzer and Gross, in preparation). Less than 1% of the total collagen is extractable from normal embryo tissues, whereas up to 80% can be dissolved from bone, skin, and aorta 48 hours after administering the drug. In addition the electron microscope shows an obvious disturbance in collagen fibrillar organization in the tissues (van den Hooff *et al.*, 1959). Levene (1961) has examined a large number of compounds for lathyrogenic activity.

Similar solubility changes can be found in the collagen of βAPN-treated guinea pigs both young and old (Gross and Levene, 1959). Lathyritic extractable collagen appears normal in terms of amino acid composition, molecular dimensions, helical conformation, and stability to thermal denaturation (Gross, unpublished). Similar to normal collagen in solution, the lathyritic collagen can be readily polymerized to fibrils by warming the solution to body temperature. However, it differs from the normal in that the fibrils redissolve upon cooling, even after incubation at body temperature for periods beyond 2 weeks (Gross, 1963) (Fig. 8). The normal collagen, in

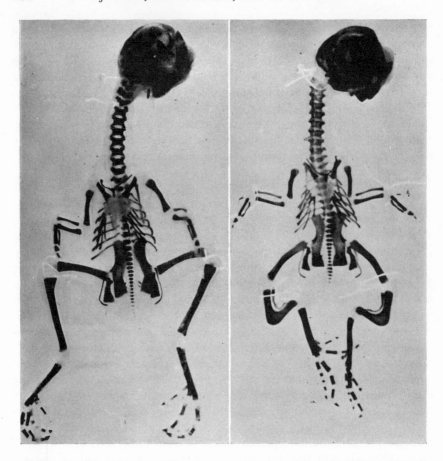

FIG. 6. Skeletons of 17-day chick embryos, normal to the left and lathyritic to the right The latter had been injected with 0.4 mg βAPN fumarate at 4 days of incubation. (From Levene and Gross, 1959; courtesy of *J. Exptl. Med.*)

contrast, becomes insoluble within 24 hours. At the intramolecular level, there is a marked increase in the proportion of noncross-linked (α type) collagen molecules in lathyritic animals (Martin *et al.*, 1961). Whether this represents a block in the normal maturation process or a breakage of pre-formed intramolecular cross links is not yet determined.

Balance studies and morphologic examination suggest that lathyritic extractable collagen is obtained from pre-existing insoluble fibrils (Levene and Gross, 1959; van den Hooff *et al.*, 1959). However, it has been proposed by others that βAPN directly or indirectly interferes with fibril formation of

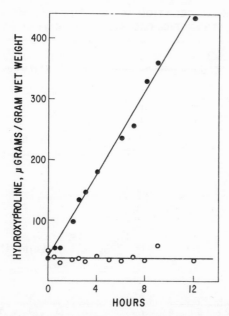

FIG. 7. Effect of administration of 20 mg βAPN fumarate on the extractability of collagen from long bones. Open circles, normal control; solid circles, lathyritic. (Tanzer and Gross, unpublished.)

newly synthesized collagen, producing an accumulation of this cold salt-extractable fraction (Follis and Tousimis, 1958; Smiley *et al.*, 1962; Smith and and Shuster, 1962; Wirtschafter and Bentley, 1962; Levene, 1962).

As a result of a recent series of isotope incorporation studies we concluded (Tanzer and Gross, manuscript in preparation) that in lathyrism, there is no obvious interference in the pathway between newly synthesized collagen and insoluble fibrils.

In one experiment H^3-proline was given 24 hours prior to the administration of βAPN. Within 24 hours after giving labeled proline most of the label was found in the insoluble fibrils, a small amount remaining free or in the very small extractable collagen pool. After administration of βAPN there followed an immediate progressive increase in extractable labeled collagen in the lathyritic, contrasting with no label accumulation in the controls (Fig. 9). This observation suggested that the insoluble collagen derived from the labeled extractable fraction during the preceding 24 hours had been returned to an extractable state. There remains the possibility that the slightly labeled collagen still being synthesized 24 hours after administration

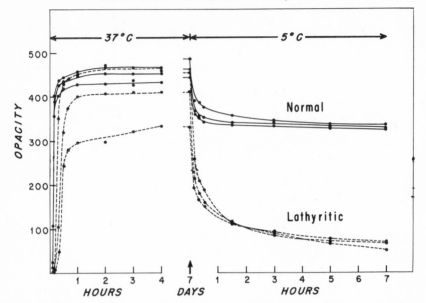

THERMAL GELATION of NaCl PURIFIED NORMAL and
LATHYRITIC COLLAGEN   (7 days at 37° C)

FIG. 8. Thermal gelation of NaCl purified normal and lathyritic collagen. Reconstitution and reversal of normal and lathyritic guinea pig collagen (three animals in each group). The six solutions were warmed to 37° and the rapid increase in opacity accompanying gelation and fibril formation is recorded on the ordinate in arbitrary units. These gels were incubated at 37° for 7 days, then cooled in an ice bath. The lathyritic collagen gels redissolved promptly, in contrast with the normal.

of H³-proline might be maintained in an extractable state, *although in fibrillar form,* by the action of βAPN. Efforts are being made to settle this question.

At the present time we do not understand how the lathyrogenic compounds act. βAPN itself and tissue fluid from lathyritic embryos do not transform collagen fibrils to an extractable state *in vitro.* There is no evidence of direct binding of the compound to the solubilized collagen in lathyritic embryos. Less than 1 mole of C¹⁴- and H³-labeled βAPN per 50 moles of protein was found in the extractable collagen of chick embryos administered the labeled compounds (Orloff and Gross, in press).

Our explanation for the source of lathyritic soluble collagen led us to reconsider D. S. Jackson's (1957) original suggestion concerning the normal pathway of collagen degradation; that is, intermolecular bonding within the fibrils is first loosened, resulting in the reappearance of extractable collagen,

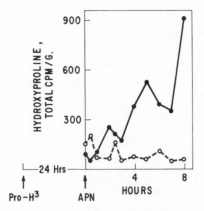

Fig. 9. Radioactive collagen accumulation in extractable collagen fraction in lathyritic embryo bone. βAPN administered 24 hours after H³-proline. Open circles, normal control; solid circles, lathyritic. (Tanzer and Gross, unpublished.)

which is then the substrate for enzymatic breakdown to amino acids. Thus, here may be two soluble collagen pools in the normal metabolic pathway, the first being newly synthesized protein on the way to form fibrils, the second soluble pool representing a step in the degradation of fibrils. One must postulate an enzymatic system which breaks intermolecular bonds to account for increased solubility, and also intramolecular bonds to account for the lack of β components in the lathyritic collagen. Jackson (1953) reported that crude hyaluronidase could cause insoluble fibrils to become soluble in dilute acid and Highberger, Gross, and Schmitt found (unpublished) that an unidentified pancreatic enzyme, as well as "erepsin," and papain could do the same thing. Thus, there is an experimental precedent for this process. Still to be accounted for is the inability of lathyritic collagen fibrils to become insoluble *in vitro*. Some subtle change in molecular structure other than that involving intramolecular bonds or the intervention of another substance must have occurred.

The malformations of the skeleton and the structural failure of blood vessels and fascia in lathyrism might well be ascribed to diminished tensile strength of collagen, this in turn resulting from an increasing proportion of collagen fibrils without firm intermolecular cross links. It is perhaps worth observing that one of the known lathyrogenic compounds, β-mercaptoethylamine, is broadly distributed throughout the tissues as part of the molecule of coenzyme A.

Collagen Degradation in Animal Tissues

It is reasonable to expect that an enzyme system must be operating in animal tissues to catabolize collagen. However, until recently a true animal collagenase has not been found (see Mandl, 1961, for review). The actual mechanisms whereby structural elements are dismantled for replacement or final removal have not been elucidated in any tissue to date. Cathepsins and other intracellular hydrolases are thought to reside in organelles, the lysosomes (de Duve, 1959), which have their drawstrings untied at the appropriate time and places, releasing the lytic principles. These would, in some selective manner, attack the right bonds in the appropriate substrates, resulting in orderly dismantling of tissues, and not interfering with the continuous function of the organism.

One of the most dramatic examples of physiologic resorption of tissue under hormonal control is the transformation of tadpole to frog. Under the influence of thyroid hormone there is a rapid resorption of the tail, the gills, the opercular areas, and the intestine (Kollros, 1961; Freiden, 1961). Weber (1957) has reported an increase in "catheptic" activity of slices of whole tail tissue during metamorphosis. He has also induced resorption in organ cultures of whole tadpole tail with thyroxine (Weber, 1962).

Fibrous collagen, as a substrate for detecting and measuring specific enzymatic activity, has the unique advantage of being relatively immune to attack (under physiologic conditions) by any of the commonly known proteolytic enzymes or cathepsins. This protein therefore seems to be a near ideal guide to a specific enzyme system operating on a particular structural element during morphogenesis.

Saline extracts of tail and gill tissue homogenates from metamorphosing and non-metamorphosing tadpoles were examined for collagenolytic activity, using soluble calf skin collagen as substrate, measuring the fall in viscosity as described by Gallop et al. (1957)—with no success. Our experience was

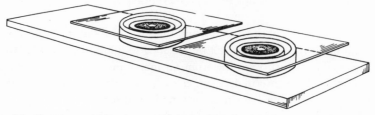

Fig. 10. Chambers mounted on glass microslides (used for collagen gel cultúres). The collagen solution is gelled within the central ring.

consistent with that of others searching for collagenase in animal tissues (Mandl, 1961; Morrione and Seifter, 1962; Woessner, 1962).

Proceeding on the assumption that appreciable amounts of unbound active collagenase may rarely, if ever, accumulate *in vivo*, a method was devised to allow the diffusion of a collagenolytic principle away from the tissue and to detect and measure its activity (Gross and Lapiere, 1962). Tissue fragments were cultivated on fibrous gels formed from calf or guinea pig skin collagen. Collagenolytic activity was detected and measured by either (1) increasing area of lysis around the explant, (2) the released hydroxyproline, or (3) C^{14} release from labeled collagen gels. It is of interest that precipitated collagen has been used as a culture substrate in recent years because it normally resists the lytic activity of growing cells, in sharp contrast to the fibrin clot (Ehrmann and Gey, 1956; Bornstein, 1958; Hillis and Bang, 1959).

Acid-extracted calf skin collagen and neutral-extracted guinea pig skin collagen heavily labeled with C^{14}-glycine, dissolved in cold neutral saline,

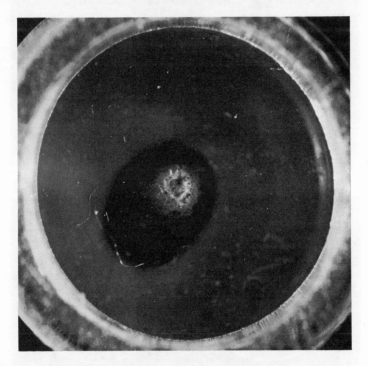

FIG. 11. Collagenolysis caused by explant of tadpole tail fin tissue. The area of lysis is the black region surrounding the tissue in the opalescent gel. (Gross and Lapiere, 1962: courtesy of *Proc. Natl. Acad. Sci. U. S.*)

were transformed to fibrous gels by warming to 37° C for several hours in chambers illustrated in Fig. 10 (Gross and Lapiere, 1962). Larger cultures were prepared in petri dishes.

It was demonstrated that reconstituted collagen substrates were immune to attack by crude tissue extracts, cathepsin C, hyaluronidase, trypsin, chymotrypsin, papain, tyrosinase, and pepsin at neutral pH. Bacterial collagenase placed on small filter paper disks at the center of the gel rapidly lysed the collagen as a linear function of time at 27 and 37° C.

Small pieces of fin tissue from *Rana catesbiana* tadpoles were applied to the surface of the collagen gels and incubated at 27 or 37° C in a moist atmosphere. Figure 11 illustrates the appearance of one of these cultures when examined by oblique illumination, 24 hours after incubation at 37° C. Figure 12A and B are photographs of cultures of rat endometrium and 2-day-

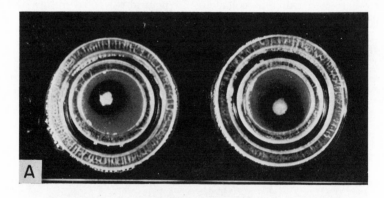

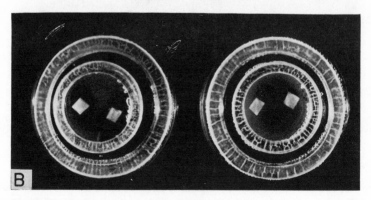

Fig. 12. A. Two cultures of rat endometrium. B. Cultures of mouse calvarium.

old rat calvarium, respectively, showing the expanding areas of collagenolysis. That the enzymatic activity could pass readily through a fine Millipore filter imposed between the tissue and the gel, is shown in Fig. 13.

Cultures prepared with frozen and thawed tadpole fin tissue displayed no lytic activity (Fig. 14), indicating that the release of enzyme is not a simple process of leaking from damaged cells.

The collagenolytic activity of a variety of non-metamorphosing tadpole tissues was tested on collagen gels in culture dishes. Figure 15 shows a set of three of these cultures before and after incubation. Of the 12 tissues tested only four proved to be active—back and tail skin, gill, and gut. These tissues undergo the most dramatic changes during metamorphosis, and also contain a high percentage of collagen. Heart, tail muscle, kidney, gonads, notochord, liver, and spinal column were inactive.

The degraded collagen was largely broken down to dialyzable peptides at neutral pH, the amount being temperature dependent. No free hydroxy-proline was produced.

By growing quantities of coarsely fragmented gill tissue on large collagen gels and allowing complete lysis, it was possible to obtain a preparation of active enzyme. A dry powder was obtained after lyophilization and assayed by the viscosometric technique and by the release of radioactivity from labeled collagen gel. Viscometry showed that the collagenolytic action of the dis-

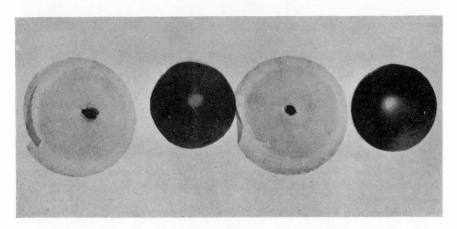

FIG. 13. Transfilter diffusion of collagenase activity from tadpole fin explant. These cultures were incubated with a fine millipore filter interposed between the collagen gel and the tissue. The fixed and stained gels are seen to the right of the filters bearing the explants. The clear hole results from collagenolytic activity. (Gross and Lapiere, 1962; courtesy of *Proc. Natl. Acad. Sci. U. S.*)

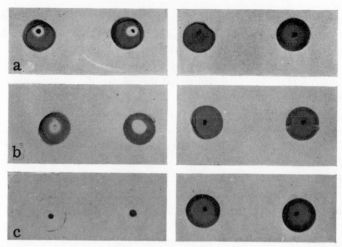

Fig. 14. Fixed and stained cultures of living (*left*) and frozen, thawed (*right*) tadpole tail fin; (a) and (b) were cultured for 48 hours and (c) for 96 hours. In (c) the collagen gel substrate has been completely lysed and only the central, rounded-up explant remains.

solved powder was biphasic, similar to bacterial collagenase. It contained approximately 0.1 unit of activity (based on bacterial collagenase standards) per mg of dried powder, low activity compared with that obtained from crude extracts of bacterial cultures. With the solid fiber assay zero order kinetics were followed, permitting a simple quantitative estimation of activity. Purification of the enzyme, characterization of its properties, and delineation of the products of collagen degradation are in progress.

The size of the tissue explant and the area of lysis proved to be directly related. The area of lysis was a linear function of the amount of collagen degraded (Fig. 18). Comparison of the rates of lysis of collagen gel cultures by bacterial collagenase and tadpole tissues revealed a rapid linear rate of activity for the former and a logarithmic rate for the latter (Fig. 17). This observation suggested that the enzyme was not available in the tissues in detectable amounts at the time of explantation, but required a lag period for its appearance, after which it was apparently present, in active form, in increasing quantities. If it were being stored in the tissues one should have observed its release from frozen and thawed explants immediately, behaving similar to the diffusion of bacterial collagenase.

It was previously reported (Gross and Lapiere, 1962) that there was little difference in collagenolytic activity of tail fin tissue between metamorphosing and non-metamorphosing tadpoles. Improvement in technique including incubating at 27° instead of 37° C, revealed significant differences. Tissues of

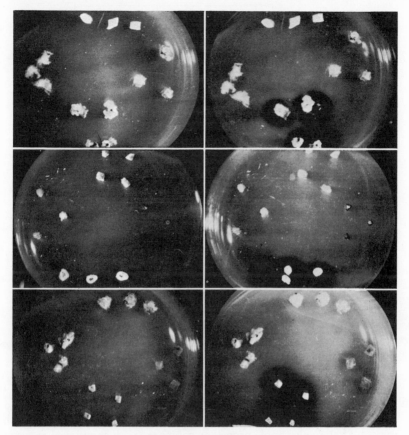

Fig. 15. Cultures of four groups of tadpole tissues in each of three dishes, 12 different tissues in all. *Left*, before incubation; *right*, after 24 hours of incubation. Lysis only in gill (*top*), gut (*middle*), and tail skin (*bottom*).

tadpoles treated with thyroxine for 6 days showed greater activity than those from untreated tadpoles (C. M. Lapiere and J. Gross, in preparation).

It is probable that tissue remodeling is controlled, in part, by manipulation of the relative rates of synthesis and degradation of structural elements. Current studies on the turnover rates of the different collagen pools in body and tail fin tissues in metamorphosing and nonmetamorphosing tadpoles are in progress. Among the changes observed was the removal of at least 50% of the water content of the tail fin and considerably less in the backskin. Assays of collagenase activity per unit area of tail fin cultured at 27° C revealed increases of two to four times that found in the same tissues of the nonmeta-

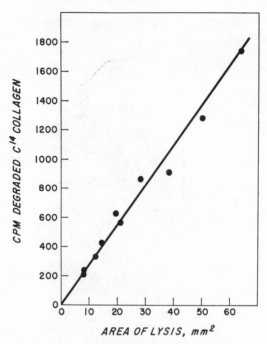

Fig. 16. Relationship between release of radioactivity and the measured area of lysis. C^{14}-labeled collagen gel used as substrate.

morphosing animals. It is quite possible that this increment is largely a result of increased concentration of cells.

The massive resorption of structures such as tail or gill during metamorphosis may be a relatively simple process as compared with complex changes taking place around the mouth parts. Localized regions of tissue probably metabolize their structural substrates at different rates with different timing, yet delicately synchronized with each other. The basic competence for resorption of collagen, for example, may be equal for all loci, but the threshold to stimulus and inhibition may vary widely. Thus the same level of thyroxine would influence each region differently. Etkin (1957) and Kollros (1961) have discussed this aspect of thyroid control of metamorphosis in some detail. Because of the repeated observations of profound differences in tissue competence in metamorphosis, i.e., tail skin transplanted to the back will resorb whereas back skin transplanted to the tail will not (Clausen, 1930), the chemical and physical characteristics of acid-extracted collagen from tail and back skin of metamorphosing and non-metamorphosing tadpoles and from the skin of young and old bullfrogs were analyzed. No differences could be found in the relative

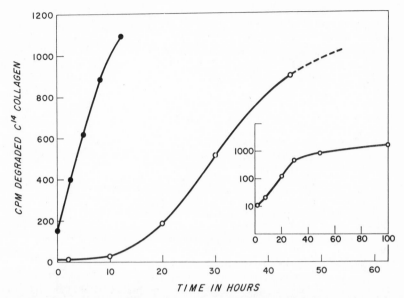

Fig. 17. Comparison of collagenolytic activity of bacterial collagenase and tadpole tissue. Solid circles, *Clostridium histolyticun* collagenase 0.004 unit; open circles, tadpole fin 3 mm². The insert shows a logarithmic plot of radioactivity released in the tissue culture. (Gross and Lapiere, 1962; courtesy of *Proc. Natl. Acad. Sci. U. S.*)

proportion of α and β subunits, or in the denaturation temperatures of these collagens.

Numerous enzyme systems (and probably reactions not directly requiring enzymes) must be involved in the over-all remodeling process. A major task will be the sorting out of those systems which have a significant physiologic role from those which are incidentally related or have only vestigial significance.

ACKNOWLEDGMENTS

Much of the work from this laboratory, described here, was done during the tenure of research grants from the National Institute of Arthritis and Metabolic Diseases, United States Public Health Service, and Eli Lilly and Company. Dr. Marvin L. Tanzer is a Postdoctoral Fellow of the Arthritis and Rheumatism Foundation.

REFERENCES

ALTGELT, K., HODGE, A. J., AND SCHMITT, F. O. (1961). Gamma tropocollagen: A reversibly denaturable collagen macromolecule. *Proc. Natl. Acad. Sci. U.S.* **47,** 1914–1924.

BEAR, R. S. (1952). The structure of collagen fibrils. *Advances in Protein Chem.* **7,** 69–160.

BLUMENFELD, O., AND GALLOP, P. M. (1962). The participation of aspartyl residues in hydroxylamine or hydrazine-sensitive bonds of collagen. *Biochemistry* **1**, 947–959.

BORNSTEIN, M. B. (1958). Reconstituted rat-tail collagen used as substrate for tissue culture on cover slips in Maximow slides and roller tubes. *Lab. Invest.* **7**, 134–137.

CHAPMAN, J. A. (1961). Morphological and chemical studies of collagen formation. *J. Biophys. Biochem. Cytol.* **9**, 639–651.

CLAUSEN, H. J. (1930). Rate of histolysis of anuran tail skin and muscle during metamorphosis. *Biol. Bull.* **59**, 199–210.

DE DUVE, C. (1959). Lysosomes: A new group of cytoplasmic particles. *In* "Subcellular Particles" (T. Hayashi, ed.), pp. 128–157. Ronald Press, New York.

DOTY, P., AND NISHIHARA, T. (1958). The molecular properties and thermal stability of soluble collagens. *In* "Recent Advances in Gelatin and Glue Research" (G. Stainsby, ed.), pp. 92–99. Pergamon Press, New York.

EDDS, M. V., JR., AND SWEENEY, P. R. (1961). Chemical and morphological differentiations of the basement lamella in molecular and cellular structure. *In* "Synthesis of Molecular and Cellular Structure," Growth Symposium No. 19 (D. Rudnick, ed.), pp. 111–138. Ronald Press, New York.

EHRMANN, R. L., AND GEY, G. O. (1956). The growth of cells on a transparent gel of reconstituted rat-tail collagen. *J. Natl. Cancer Inst.* **16**, 1375–1403.

ETKIN, W. (1955). Metamorphosis. *In* "Analysis of Development" (B. H. Willier, P. A. Weiss, and V. Hamburger, eds.), pp. 631–663. Saunders, Philadelphia, Pennsylvania.

FOLLIS, R. H., AND TOUSIMIS, A. J. (1958). Experimental lathyrism in the rat: Nature of defect in epiphysial cartilage. *Proc. Soc. Exptl. Biol. Med.* **98**, 843–848.

FRIEDEN, E. (1961). Biochemical adaptation and anuran metatamorphosis. *Am. Zoologist* **1**, 115–149.

GALLOP, P. M., SEIFTER, S., AND MEILMAN, E. (1957). Studies on collagen I. The partial purification, assay and mode of activation of bacterial collagenase. *J. Biol. Chem.* **277**, 891–906.

GALLOP, P. M., SEIFTER, S., AND MEILMAN, E. (1959). The occurrence of ester-like linkages in collagen. *Nature* **183**, 1659–1661.

GARRAULT, H. (1937). Structure de la membrane basale sous-epidermisque chez les embryons de selaciens. *Arch. anat. microscop.* **33**, 167–180.

GEIGER, B. J., STEENBOK, H., AND PARSON, H. T. (1933). Lathyrism in the rat. *J. Nutrition* **6**, 427–442.

GILLETTE, E. P. (1872). *In* "Anatomie et physiologie du tissu conjonctif ou lamineau," p. 17. Ballière, Paris.

GLIMCHER, M. J., HODGE, A. J., AND SCHMITT, F. D. (1957). Macromolecular aggregation states in relation to mineralization: A collagen hydroxyapatite system as studied *in vitro*. *Proc. Natl. Acad. Sci. U.S.* **43**, 860–867.

GOULD, B. S., AND WOESSNER, J. F. (1957). Biosynthesis of collagen. The influence of ascorbic acid on the proline, hydroxyproline, glycine and collagen content of regenerating guinea pig skin. *J. Biol. Chem.* **226**, 289–300.

GREEN, N. M., AND LOWTHER, D. A. (1959). Formation of collagen hydroxyproline *in vitro*. *Biochem. J.* **71**, 55–66.

GROSS, J. (1956). The behavior of collagen units as a model in morphogenesis. *J. Biophys. Biochem. Cytol.* **2**, 261–274.

GROSS, J. (1958). Studies on collagen formation. III. Time dependent solubility changes of collagen *in vitro*. *J. Exptl. Med.* **108**, 215–226.

GROSS, J.   (1959).   On the significance of the soluble collagens. *In* "Connective Tissue, Thrombosis and Atherosclerosis" (I. H. Page, ed.), pp. 77–95. Academic Press, New York.

GROSS, J.   (1961a).   Collagen. *Sci. American* **204,** 120–130.

GROSS, J.   (1961b).   Ageing of connective tissue; The extracellular components. *In* "Structural Aspects of Ageing" (G. Bourne, ed.), pp. 177–195. Pitman, London.

GROSS, J.   (1963).   An intermolecular defect in collagen in experimental lathyrism. *Biochim. et Biophys. Acta* (in press).

GROSS, J., AND KIRK, D.   (1958).   The heat precipitation of collagen from neutral salt solutions: Some rate-regulating factors. *J. Biol. Chem.* **233,** 355–360.

GROSS, J., AND LAPIERE, C. M.   (1962).   Collagenolytic activity in amphibian tissues: A tissue culture assay. *Proc. Natl. Acad. Sci. U.S.* **48,** 1014–1022.

GROSS, J., AND LEVENE, C. I.   (1959).   Effect of β-aminopropionitrile on extractability of collagen from skin of mature guinea pigs. *Am. J. Pathol.* **35,** 687.

GROSS, J., HIGHBERGER, J. H., AND SCHMITT, F. O.   (1952).   Some factors involved in the fibrogenesis of collagen. *Proc. Soc. Exptl. Biol. Med.* **80,** 452–465.

GROSS, J., HIGHBERGER, J. H., AND SCHMITT, F. O.   (1954).   Collagen structures considered as states of aggregation of a kinetic unit. The tropocollagen particle. *Proc. Natl. Acad. Sci. U.S.* **40,** 679–688.

GUSTAVSON, K. H.   (1956).   "The Chemistry and Reactivity of Collagen." Academic Press, New York.

HARKNESS, R. D.   (1961).   Biological functions of collagen. *Biol. Revs. Cambridge Phil. Soc.* **36,** 399–463.

HARKNESS, R. D., MARKO, A. M., MUIR, H. M., AND NEUBERGER, A.   (1954).   Metabolism of collagen and other proteins of the skin of rabbits. *Biochem. J.* **56,** 558–569.

HARRINGTON, W. F., AND VON HIPPEL, P. H.   (1961).   The structure of collagen and gelatin. *Advances in Protein Chem.* **16,** 1–138.

HAUSMANN, E., AND NEUMAN, W. F.   (1961).   Conversion of proline to hydroxyproline and its incorporation into collagen. *J. Biol. Chem.* **236,** 149–152.

HIGHBERGER, J. H., GROSS, J., AND SCHMITT, F. O.   (1951).   The interaction of mucoprotein with soluble collagen, an electron microscope study. *Proc. Natl. Acad. Sci. U.S.* **37,** 286–291.

HILLIS, W. D., AND BANG, F. B.   (1959).   Cultivation of embryonic and adult liver cells on a collagen substrate. *Exptl. Cell Research* **17,** 557–560.

HODGE, A. J.   (1960).   Principles of ordering in fibrous systems. *Proc. Intern. Conf. Electron Microscopy, 4th* **2,** 119–139.

HODGE, A. J., AND SCHMITT, F. O.   (1960).   The charge profile of the tropocollagen macromolecule and the packing arrangement in native type collagen fibrils. *Proc. Natl. Acad. Sci. U.S.* **46,** 186–197.

JACKSON, D. S.   (1953).   Chonchroitin Sulfuric Acid as a factor in the stability of Tendon. *Biochem. J.* **54,** 638.

JACKSON, D. S.   (1957).   Connective tissue growth stimulated by carrageenin. 1. The formation and removal of collagen. *Biochem. J.* **65,** 277–284.

JACKSON, D. S., AND BENTLEY, J. P.   (1960).   On the significance of the extractable collagens. *J. Biophys. Biochem. Cytol.* **7,** 37–42.

JACKSON, D. S., AND FESSLER, T. H.   (1955).   Isolation and properties of a collagen soluble in salt solution at neutral pH. *Nature* **176,** 69–70.

JAKUS, M. A. (1956). Studies on the cornea. II. Fine structure of Decemet's membrane. *J. Biophys. Biochem. Cytol.* **2**, 243–252.

JAKUS, M. A. (1960). Fine structure of certain ocular tissues. *Proc. Intern. Conf. Electron Microscopy, 4th* **11**, 344–347.

KARRER, H. E. (1960). Electron microscope study of developing chick embryo aorta. *J. Ultrastruct. Research* **4**, 420–454.

KEMP, N. E. (1961). Replacement of the larval basement membrane lamella by adult-type basement membrane in anuran skin during metamorphosis. *Develop. Biol.* **3**, 391–410.

KOLLROS, J. J. (1961). Mechanisms of amphibian metamorphosis: Hormones. *Am. Zool.* **1**, 107–114.

LEVENE, C. I. (1962). Studies on the mode of action of lathyrogenic compounds. *J. Exptl. Med.* **116**, 119–130.

LEVENE, C. I. (1961). Structural requirements for lathyrogenic agents. *J. Exptl. Med.* **114**, 295–310.

LEVENE, C. I., AND GROSS, J. (1959). Alterations in the state of molecular aggregation of collagen induced in chick embryo by β-aminopropionitrile (Lathyrus Factor). *J. Exptl. Med.* **110**, 771–790.

LOWTHER, D. A., GREEN, N. M., AND CHAPMAN, J. A. (1961). Morphological and chemical studies of collagen formation. *J. Biophys. Biochem. Cytol.* **10**, 373–388.

MANDL, I. (1961). Collagenases and elastases. *Advances in Enzymology* **23**, 163–264.

MANNER, G., AND GOULD, B. S. (1962). Collagen biosynthesis. The formation of an S-RNA-hydroxyproline complex. *Federation Proc.* **21**, 169d.

MARTIN, G. R., GROSS, J., PIEZ, K. A., AND LEWIS, M. S. (1961). On the intramolecular crosslinking of collagen in lathyritic rats. *Biochim. et Biophys. Acta* **53**, 599–601.

MIZUHIRA, V. (1951). On mechanism of formation of cutis-fibres of amphibian larvae (A contribution to development of connective tissue fibres). *Arch. hist. jap. (Okayama)* **2**, 445–463.

MORRIONE, T. G., AND SEIFTER, S. (1962). Alterations in the collagen content of the human uterus during pregnancy and post partum involution. *J. Exptl. Med.* **115**, 357–65.

NAGEOTTE, J. (1927). Sur le caillot artificel de collagène: signification, morphologie, générale et technique. *Compt. rend. soc. biol.* **96**, 172–174.

NIIZIMA, M., MIZUHIRA, V., TAKAGI, A., AND OKADA, S. (1954). Comparative anatomical studies on the arrangement of connective tissue fibres of the corium of vertebrates. *Bull. Tokyo Med. and Dental Univ.* **1**, 15–20.

NEUBERGER, A., AND SLACK, H. G. B. (1953). The metabolism of collagen from liver, bone, skin and tendon in the normal rat. *Biochem. J.* **53**, 47–52.

OREKHOVITCH, V. N., AND SHPIKITER, V. O. (1955). *Doklady Akad. Nauk S.S.S.R.* **101**, 529.

OREKHOVITCH, V. N., SHPIKITER, V. O., KAZAKOVA, O. V., AND MAZOUROV, V. I. (1959). The incorporation of C^{14} labeled glycine into the α and β components of procollagen. *Arch. Biochem. Biophys.* **85**, 554–556.

ORLOFF, S. D., AND GROSS, J. (1963). Experimental lathyrism in the chick embryo. The distribution of β-amino propionitrile. *J. Exptl. Med.* (in press).

PETERKOFSKY, B., AND UDENFRIEND, S. (1961). Conversion of proline[14] to peptide-bound hydroxyproline-C^{14} in a cell-free system from chick embryo. *Biochem. Biophys. Research Communs.* **6**, 184–190.

PIEZ, K. A., AND LIKINS, R. C. (1957). The conversion of lysine to hydroxylysine and its relation to the biosynthesis of collagen in several tissues of the rat. *J. Biol. Chem.* **299**, 101–109.

PIEZ, K. A., LEWIS, M. S., MARTIN, G. R., AND GROSS, J. (1961). Subunits of the collagen molecule. *Biochim. et Biophys. Acta* **53**, 596–598.

PONSETI, I. V., AND SHEPARD, R. S. (1954). Lesions of the skeleton and of other mesodermal tissues in rats fed sweet pea seeds. (*Lathyrus oderatous*). *J. Bone and Joint Surg.* **36A**, 1031–1058.

PORTER, K. R., AND PAPPAS, G. D. (1959). Collagen formation by fibroblasts of chick embryo dermis. *J. Biophys. Biochem. Cytol.* **5**, 143–166.

PROCKOP, D. J., PETERKOFSKY, B., AND UDENFRIEND, S. (1962). Studies on the intracellular localization of collagen synthesis in the intact chick embryo. *J. Biol. Chem.* **237**, 1581–1584.

RANDALL, J. T., ed. (1953). "Nature and Structure of Collagen." Butterworths, London.

RICH, A., AND CRICK, F. H. C. (1961). The molecular structure of collagen. *J. Mol. Biol.* **3**, 483–506.

ROBERTSON, W. VON B., HEWITT, S., AND HERMAN, C. (1959). The relation of the conversion of proline into hydroxyproline in the synthesis of collagen in the carrageenan granuloma. *J. Biol. Chem.* **234**, 105–108.

SCHMITT, F. O. (1959). Interaction properties of elongate protein macromolecules with particular reference to collagen (tropocollagen). *Revs. Modern Phys.* **31**, 349–358.

SCHMITT, F. O. (1960). Contributions of molecular biology to medicine. *Bull. N.Y. Acad. Med.* **36**, 725–749.

SCHMITT, F. O., HALL, C. E., AND JAKUS, M. A. (1942). Electron microscope investigations of the structure of collagen. *J. Cellular Comp. Physiol.* **20**, 11–33.

SCHMITT, F. O., GROSS, J., AND HIGHBERGER, J. H. (1953). A new particle type in certain connective tissue extracts. *Proc. Natl. Acad. Sci. U.S.* **39**, 459–470.

SELYE, H. (1957). Lathyrism. *Rev. can. biol.* **16**, 1–82.

SINEX, S. M., VAN SLYKE, D. D., AND CHRISTMAN, D. R. (1959). The source and state of the hydroxylysine of collagen. II. Failure of free hydroxylysine to serve as a source of the hydroxylysine or lysine of collagen. *J. Biol. Chem.* **234**, 918–921.

SMILEY, J. D., YEAGER, H., AND ZIFF, M. (1962). Collagen metabolism in osteolathyrism in chick embryos. Site of action of β-aminopropionitrile. *J. Exptl. Med.* **116**, 45–54.

SMITH, D. J., AND SHUSTER, R. C. (1962). Biochemistry of lathyrism. I. Collagen biosynthesis in normal and lathyritic chick embryos. *Arch. Biochem. Biophys.* **98**, 418–50.

SMITH, D. J., AND TORTERELLA, J. B. (1962). Incorporation of radioactive sulfate by bone and normal lathyritic chick embryos. *Federation Proc.* **21**, 172a.

STAINSBY, B. Y. G., ed. (1958). "Recent Advances in Gelatin and Glue Research." Pergamon Press, New York.

STEARNS, M. L. (1940). Studies on the development of connective tissue in transparent chambers in rats ear. I. *Am. J. Anat.* **66**, 133–176; **67**, 55–97.

STETTEN, M. R. (1949). Some aspects of the metabolism of hydroxyproline studied with the aid of isotopic nitrogen. *J. Biol. Chem.* **181**, 31–37.

TANZER, M. L., AND GROSS, J. Manuscript in preparation.

THOMPSON, R. C., AND BALLOU, J. E. (1956). Studies of metabolic turnover with tritium with a tracer. V. The predominantly non-dynamic state of body constituents in the rat. *J. Biol. Chem.* **223**, 795–809.

TOMKINS, G. M., AND YIELDING, K. L. (1961). Regulation of the enzymic activity of glutamic dehydrogenase mediated by changes in its structure. *Cold Spring Harbor Symposia Quant. Biol.* **26**, 331–341.

TUNBRIDGE, R. E., ed. (1956). "Connective Tissue, A Symposium." Blackwell, Oxford.

VAN DEN HOOFF, A., LEVENE, C. I., AND GROSS, J. (1959). Morphologic evidence for collagen changes in chick embryos treated with β-aminopropionitrile. *J. Exptl. Med.* **110**, 1017–1022.

WARNER, J. R., RICH, A., AND HALL, C. E. (1962). Electron microscope studies of ribosomal structures synthesizing hemoglobin. *Science* **138**, 1399–1403.

WEBER, R. (1957). On the biological function of cathepsin in tail tissue of *Xenopus* larvae. *Experientia* **13**, 153–155.

WEBER, R. (1962). Induced metamorphosis in isolated tails of *Xenopus* larvae. *Experientia* **18**, 84.

WEISS, P. (1957). Macromolecular fabrics and patterns. *J. Cellular Comp. Physiol.* **49**, 105–112.

WEISS, P., AND FERRIS, W. (1956). The basement lamella of amphibian skin, its reconstruction after wounding. *J. Biophys. Biochem. Cytol.* **2**, 275–282.

WIRTSCHAFTER, Z. T., AND BENTLEY, J. P. (1962). The extractable collagen of lathyritic rats with relation to age. *Lab. Invest.* **11**, 365–367.

WOESSNER, J. F. (1962). Catabolism of collagen and non-collagen protein in the rat uterus during post-partum involution. *Biochem. J.* **83**, 304–314.

YARDLEY, J. H., HEATON, M. W., GAINES, L. M., AND SHULMAN, L. E. (1960). Collagen formation by fibroblasts. Preliminary electron microscopic observations using thin sections of tissue culture. *Bull. Johns Hopkins Hosp.* **106**, 381–393.

On Mechanisms of Elongation

PAUL B. GREEN

Division of Biology, University of Pennsylvania, Philadelphia, Pennsylvania

"As we shall see below, tool and product are here in a reciprocal relation, which is perhaps most simply illustrated by the grooving action of water flowing downhill—the river fashioning its bed and the bed confining the river. Processes oriented in a given direction . . . can result in structural orientation, which then, in turn, will channel further flow" (Weiss, 1962).

Morphogenesis is that branch of developmental biology concerned with phenomena governing the shape of cells, tissues, and organisms. During development new organs appear in specific regions, and if the organ, be it a leaf or a limb, has a major axis, this axis will lie in a specific direction with reference to the body axes. This general phenomenon, control of geometric relations during growth, is manifested in simplest form in elongation; here new synthesis or cell rearrangement is carried out to extend a single axis of an enlarging structure. Despite the relative simplicity, the mechanisms involved are diverse and often difficult to reduce to cellular, much less molecular, terms. While the ultimate subject of this paper will be the elongation of a single plant cell, this topic will be approached through a consideration of elongation in general.

Two meanings of elongation. In development the word elongation can mean either the *transition* from a round form to an elongate one, as seen after neurulation in the frog embryo, in rhizoid formation in the *Fucus* egg, etc., or it can refer to the mechanisms involved in the *extension* of a pre-existing long axis. Elongation in the latter sense has more the character of a steady-state process than a unique event and will receive emphasis here.

The resolution of elongation of the organism into events at the tissue and cellular level is considerably more direct in higher plants and some invertebrates than in vertebrates. In the larval vertebrate, the tissue primarily involved in elongation is believed to be the notochord because excision of segments of this structure leads to stunted embryos (Kitchin, 1949). Further, the notochord is capable of elongating by itself in culture (Holtfreter, 1939). Examination of notochord elongation reveals that it involves the cellular activities of proliferation, migration, assumption of a disklike form normal to the notochord axis, and subsequent inflation (hypertrophy) (Mookerjee *et al.*, 1953). A directing action of the notochord sheath may also be present (Mookerjee, 1953). The appearance of an amphibian tail bud suggests that the same cellular activities combine to bring about the prolonged extension of the tail. The most obviously directed feature of the elongation is the assemblage of flat cells normal to the notochord axis.

Grosser vertebrate elongation takes place in the vertebrae and long bones (Fig. 1). In the latter an essential feature again appears to be the alignment of a stack of disk-shaped cells normal to the axis of elongation; the cartilage cells so aligned then hypertrophy to a near-spherical form to extend the organ (Dodds, 1930). This reveals that much of the control of the direction of growth takes place before, and presumably independently of, the cell inflation process. This latter hypertrophy could be in itself undirected beyond the fact that it takes place in a disk-shaped cell. The roundabout nature of the cellular basis for bone elongation is striking when compared with elongation in certain invertebrate forms or in plant systems, as shown in Fig. 1.

A simple type of elongation in larvae of annelids, molluscs, and some arthropods appears to be causally related to the production of long files of daughter cells from individual teloblast cells (E. B. Wilson, 1889). The teloblasts are located in the posterior part of the embryo and are large cells which bud off their progeny at the anterior end of the cell. The alternate enlargement and division of the teloblast in a fixed direction appears to be the essential feature of elongation, but to the writer's knowledge has not been investigated with regard to subcellular details. The same may be said for the apparently simple elongation of the rootlike stolon of attached coelenterates such as *Obelia*. Here the elongating tip is a single hollow structure just two cell layers thick (Berrill, 1961). The cells of each layer proliferate (interstitial cells are absent) in a predictable location and orientation. Interdigitation of the progeny, however, must occur to bring the daughter cell into the columnar epithelial configuration (Berrill, 1949).

In comparison with the analogous problem in animals, considerable progress has been made toward understanding the detailed mechanism of elongation of gross plant organs such as stems, monocotyledonous leaves, coleoptiles, and roots. Because of their cell wall, plant cells cannot display locomotion and the essential feature of plant elongation is therefore the directed expansion of the cell. In the cylindrical portion of the root, for example, the cells toward the tip both elongate and divide to produce new length in the form of new cells. Farther back, elongation occurs alone and produces new length in the form of longer cells. It is unfortunate that the two districts are often termed, respectively, "zone of cell division" and "zone of cell elongation" because elongation—in the sense of the separation

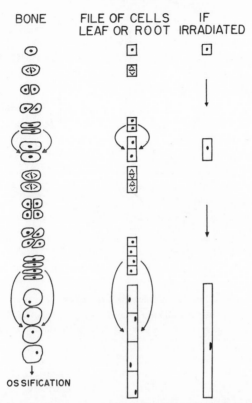

FIG. 1. Diagram of cartilage cell behavior characteristic of elongation in bone (after Dodds, 1930), growth of typical plant tissue, and gamma-irradiated wheat seedlings (after Foard and Haber, 1961). Arrows show the steps which actually elongate the structure.

of points along a line—is common to both zones (Erickson and Sax, 1956a,b). The difference between the two zones is cell size, not the presence or absence of elongation. The persistence of elongation in proximal regions suggests that the directed expansion of the cells is independent of the other highly ordered cell structure in elongating organs, the mitotic spindle. This view is borne out in the experimental work of Foard and Haber (1961) and Haber (1962) on wheat leaf growth. In the normal seedling leaf, divisions and elongation are coupled so that long files of cells are formed. When the seeds receive large doses of gamma radiation, the divisions do not occur but the leaf still extends manyfold. The typical files of cells are replaced by extremely elongated single cells (Fig. 1). Thus in the elongation process in single plant cells, an essential feature of plant organ elongation has been retained. This

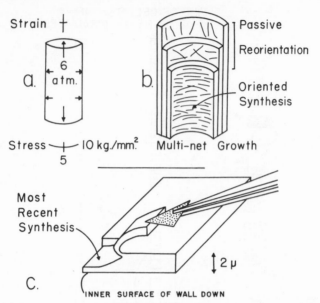

FIG. 2. (a) Relation between wall stresses and wall growth in the plant cell. (b) Micro-fibrillar arrangement in growing cell walls. (c) Method of obtaining samples of recently built cell wall.

generalization applies to cells with a diffuse distribution of expansion along the surface because the tissue cells of plants have this feature (Castle, 1955; K. Wilson, 1957). Tip-growing cells such as root hairs, pollen tubes, and some cambial derivatives do not contribute greatly to the elongation of the plant. For this reason they will not be discussed here; see books by Roelofsen (1959) and Frey-Wyssling (1959). In this connection it is perhaps interesting to note that the most conspicuous tip-elongating animal cell, the neuron, likewise does not contribute directly to elongation of organs (see Weiss, 1955, 1959).

The point has been made that the elongation of cylindrical plant organs such as shoots and roots directly reflects the elongation of single cells. This generalization applies most clearly to the cylindrical part of these organs. Additional complexities are found in other parts such as the rounded tip of the root, where new files of cortical cells are added by special divisions (Heimsch, 1960).

The simplest self-contained elongating plant systems are single cells or uniseriate filaments such as *Spirogyra*. We have chosen to study the largest of all exposed and growing cylindrical cell surfaces, that of the single-celled internode of *Nitella* (Fig. 3).

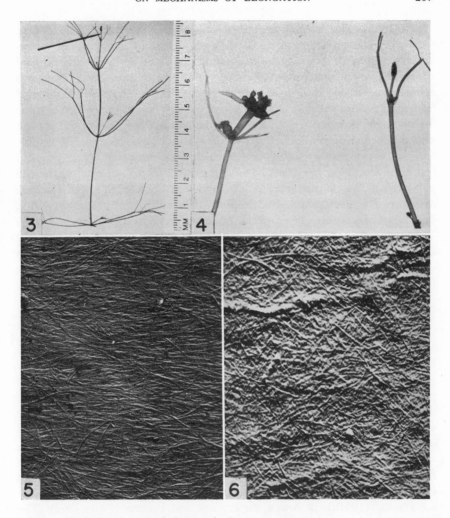

FIG. 3. Typical experimental plant, scale shown.

FIG. 4. *Left*, a shoot grown in 0.35% colchicine. *Right*, a control. 2×.

FIG. 5. Electron micrograph of a shadowed replica of a normally growing *Nitella* internode cell wall. Note predominant transverse orientation of microfibrils. Cell axis is parallel to the long side of the page. 18,000×.

FIG. 6. A similar photograph but of a wall that had been increasing in mass while in the drug colchicine. The cell was physically restrained from expanding during treatment. Note the apparently random orientation of microfibrils. 14,000×.

Nitella is a large fresh-water alga which grows in streams and also may form extensive tangled mats on the bottom of shallow ponds. When grown in the laboratory in an autoclaved mixture of garden soil (2%) and distilled water (98%), the plants may attain a growth rate of 2 cm/day. The plants consist of nodes and internodes, the internode being a single cell. While first formed as a short cylinder only 30 μ long, it will typically elongate to some 3 cm. Single cells may elongate 1 cm/day toward the end of the growth period. The elongation occurs evenly along the cell axis (Green, 1954). The cell has a remarkably strong predisposition to elongate. When physical restraints are placed on a cell to obtain growth in new directions, the cell will often slip out of the restraining structure or even break it and elongate as before. The direction of cylindrical growth is influenced by light or gravity, but the *cylindrical nature* of growth is extremely hard to alter experimentally. This behavior stands in apparent contrast to the elongation of the animal cell which displays many reversible changes in form and whose elongation can be directly influenced by relatively subtle textural features of the milieu (Weiss, 1959).

Axis Extension Related to Ordered Synthesis of Cell Wall

Theory of Plant Cell Growth

A general theory for plant cell expansion proposed in the early 1930's (reviews by Heyn, 1940 and Ray, 1961) has stood the test of time remarkably well. The theory states that the cell expands because the strong peripheral part of the cell, the cell wall, gradually yields to the hydrostatic pressure of the cell vacuole (Fig. 2a). The pressure developed in the vacuole is usually about 6 atmospheres (Guillard, 1962). The slowly yielding wall is comprised of very long microfibrils of crystalline carbohydrate, such as cellulose, embedded in an amorphous carbohydrate matrix. In *Nitella opaca* the microfibrils are of cellulose I (B) (Probine and Preston, 1961). For details on wall chemistry of higher plants see Northcote (1958) and Jensen (1961); of algae see Preston (1961a); of *Nitella* see Probine and Preston (1961, 1962). The microfibrils always lie in the plane of the cell wall (Fig. 2b). The wall is thus a two-phase system comparable with the glass rod and plastic structure of "Fiberglas," one of the strongest materials, on a weight basis, known to man (Slayter, 1962). Dimensions are such that the growing cell wall is under stresses of 10's of kg/mm^2 (Green, 1962).

While the above account suggests a rather simple inflation process, consideration of the magnitude of the expansion makes it clear that the protoplast must actively maintain the osmotic pressure of the vacuole (primarily through the uptake of inorganic ions from the environment) and must synthesize new wall materials to compensate for the thinning tendency during expansion. Many algal cells (*Valonia, Nitella, Acetabularia*) increase their surface area many thousandfold during development, and cell wall thickness may remain constant throughout or may even increase two- or threefold (Green, 1958b). Wall synthesis is characteristic of growth.

The site of the addition of new wall material has been shown in a variety of cases to be at or near the inner surface of the cell wall. This had long been suspected from the lamellar appearance of certain growing algal cells (see Küster, 1956; Mühlethaler, 1961), most especially the giant cell of *Valonia* where the wall structure is continuous near the protoplast, but appears to be torn and in the process of sloughing off in the peripheral part (Steward and Mühlethaler, 1953). The pattern of microfibrillar orientation in cylindrical cells strongly suggests growth of the wall at its inner surface (Houwink and Roelofsen, 1954; Roelofsen, 1958). This mode of addition of new substance, apposition, has been shown to be characteristic of the *Nitella* internode cell. Walls growing in the presence of tritiated water incorporated the label at or near the inner surface of the wall, as could be shown by the relative radioactivity of the inner and outer wall surface (Green, 1958a). An apparent exception to the generalization of wall growth by apposition is the outer epidermal wall of the *Avena* coleoptile, where some new mass is added in the wall interior by intussusception (Setterfield and Bayley, 1959).

Anisotropy of Growth

If the above mechanism for growth, namely the yielding of the wall to the vacuole pressure, is correct, then any *directed* features of cell expansion must have a basis in the properties of the cell wall. Hydrostatic pressure is the same in all directions and directed growth has to be controlled by the unequalness of yield (physical anisotropy) of the cell wall. The wall should yield relatively easily in the major direction of growth (longitudinally) and should be very strong in the transverse direction. Such physical properties have been found for the slow yield ("creep") of the *Nitella* internode cell by Probine and Preston (1961, 1962). This *physical anisotropy* had long been assumed because of remarkable *optical anisotropy* of growing cylindrical cell walls. In the polarized light microscope cylindrical walls are typically birefringent with the high index of refraction running at right angles to the cell axis (Frey-Wyssling, 1959). Because it was known that the high index of refraction of cellulose runs parallel to the axis of the cellulose molecule, it was deduced that in most growing cell walls the cellulose—or microfibrillar—component must run transversely, at right angles to the direction of maximum extension. This deduction was proven correct by the direct observation with the electron microscope of microfibrils in the cell wall (Houwink and Roelofsen, 1954; Wardrop, 1956). The microfibrils apparently give added strength to the wall in its transverse direction to prevent bulging and permit extension.

The above picture applies to a wide variety of higher plant cell types (root cells, coleoptile parenchyma, etc.) and to the internode of *Nitella* where cells of all lengths have a predominantly transverse microfibrillar orientation, as seen in Fig. 5 (Green, 1958b).

Besides the above correlation between transverse texture and elongation, there is some circumstantial experimental evidence that wall anisotropy controls the axial nature of expansion. Cells treated with high concentrations of auxin, the plant growth hormone, tend to grow unusually fat, being shaped like barrels (Diehl *et al.*, 1939) or relatively large spheres (tobacco pith, Das *et al.*, 1956; fern gametophytes, Miller, 1961). Auxin normally accelerates elongation by increasing the plasticity of the wall (review by Setterfield and Bayley, 1961). A possible explanation of the above formative effects is that excessive plasticity leads to a breakdown of the transverse order in the wall, so that control of the polarity of growth is lost. Unfortunately the microfibrillar arrangement in the abnormal cells was not determined.

There are some exceptions to the above broad correlation between transverse wall texture and elongation. The outer epidermal wall of the coleoptile has a net longitudinal microfibrillar orientation and various algae have a unique crisscross pattern of microfibrils in the wall (see Fig. 19, also Preston, 1952). Further, the elongate cells of the red alga *Griffithsia* have a very nearly random dispersion of microfibrils in the plane of the wall (Myers *et al.*, 1956). The above exceptions require that further evidence be given to show that the conventional transverse microfibrillar arrangement in *Nitella* is significant in the control of form.

Significance of Transverse Wall Texture in Elongation in *Nitella*

One test of the role of transverse wall texture in cell morphogenesis involves the use of the drug colchicine. When grown in this compound small cylindrical *Nitella* internodes grow into spheres or fat cylinders (Figs. 4, 7, and 8). This was reported by Delay (1957) for the closely related genus *Chara* and was easily repeated in *Nitella*, a concentration of 0.35% being effective. We further observed that the wall texture of the spherical cell was isotropic or random in the plane of the wall (Fig. 6). This randomness is evident in the living cell in polarized light (Figs. 7 and 8). Thus loss of organized wall texture and loss of directed growth are correlated.

A second piece of evidence involves the formation of short laterals from the central part of the cell. When a young cell is confined in a jacket of dialysis tubing with a hole in it, a protrusion will occasionally form and display elongation at right angles to that of the rest of the cell (Fig. 9). The wall structure of the lateral is transverse to its own axis (Fig. 10), again providing

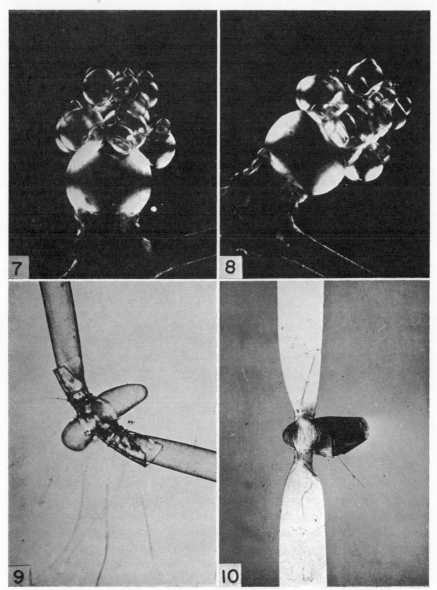

FIGS. 7 AND 8. Two views of a small living shoot that had grown in colchicine. The pictures were taken in polarized light (crossed prisms) with a 45° rotation between exposures. Persistence of the black cross in the large spherical cell reveals isotropy (randomness) of wall texture. The fibrils, randomly arranged in the plane of the wall, give brightness at the periphery of the cross because they are not perpendicular to the beam of polarized light. 20×.

FIG. 9. A large and small lateral growing out of holes in a dialysis-tubing jacket. The jacket was held on by strands of nylon thread. 16×.

FIG. 10. Same preparation, cleaned and flattened, seen in polarized light with the compensator turned. Areas darker than the background have orientation (high index of refraction) of the wall parallel to the long side of the page, areas lighter parallel to the base of the page. Transverse order in mother wall and the new lateral. 17×.

support for the view that wall texture and directed expansion are intimately related.

With confidence in the significance of transverse synthesis for elongation in *Nitella*, we undertook to determine to what extent the ordering device, whatever its nature, gained its geometrical information from the pre-existing wall texture. We wanted to test whether transverse synthesis was based on crystallization in conformity with recently made material. This is a widespread mechanism for the perpetuation of order. It occurs in the longitudinal and lateral extension of striated muscle, in the various patterns of aggregation of the tropocollagen molecule (Schmitt, 1959), and in the repair of the amphibian basement membrane where collagenous fibrils have their directions "determined by the stumps of the old fibers at the former wound edge" (Weiss, 1962). If, on the other hand, the direction of new microfibrils could be shown to be *independent* of pre-existing wall texture, it would indicate that ordered synthetic machinery, not ordered product, was significant. This appears to be the case.

The experiment involved use of colchicine to induce the production of a layer of randomly arranged microfibrils which, upon removal of the drug, was covered up by transversely ordered microfibrils. The experiment was done in such a way that the thickness of the random layer could be measured and the quality of the superimposed transverse layer could be compared with the normal wall in terms of optical properties.

To present the experiment a few comments on the use of polarized light and the peculiarities of its application to cell walls must be made.

Polarized Light Microscopy

The presence of microfibrils with a preferred orientation in a wall is detected by the wall appearing bright when viewed between crossed polarizing prisms. The brightness is at a maximum when the fibrils run at 45° to the planes of transmission of the prisms. If a wall appears bright when its axis is at the 45° position, it means that the microfibrils run either generally parallel to the cell wall axis or perpendicular to it. This ambiguity is resolved by the insertion into the beam of a crystal of known birefringent properties. As this crystal or compensator is rotated in the plane normal to the beam, it will alternately add to the action of the specimen and subtract from it. In the subtraction position the specimen appears dark ("compensated") and this position can be readily found. From knowledge of the exact position of the compensator crystal the direction of the high index of refraction, and thus the direction of the microfibrils of the specimen, can be directly deduced. If the high index of refraction runs parallel to the long axis of the object, as in the A-band of muscle, the object is termed positively birefringent. If it runs at right angles to the axis of the object, as in the *Nitella* cell wall, the object is negatively birefringent. The action of the specimen on the polarized beam is called retardation; it is a distance between 2 wave fronts as they emerge from the specimen. Retardation (Γ) is a function of the molecular order within the specimen, in this case the degree of alignment of the microfibrils, and specimen thickness. Retardation per unit thickness is a pure number called birefringence. Retardation is measured with a rotating compensator (the more turning needed to compensate, the greater the retardation); thickness is measured in a separate interference microscope. Measurement of both polarized light retardation and thickness of cell walls gives considerable information on changes in wall texture. Thus if in comparing young and old walls there had been an increase in retardation of 50% but an increase in wall thickness of 75%, there has occurred a decrease in overall transverse order in the wall despite the greater retardation (greater brightness). An excellent account of polarized light microscopy is found in an article by Bennet (1950); quantitative interference microscopy is well described by Hale (1960).

Peculiar Optical Properties of Elongating Cell Walls

In studying the nature of newly deposited materials in elongating cell walls, there is the basic difficulty that recently made microfibrils do not stay put. The transverse fibrils tend to become passively reoriented by elongation as they are covered up by new depositions (Fig. 2b). This rearrangement prohibits the calculation of the order in new synthesis by subtraction of measurements in thin (young) walls from values of thick (old or treated) walls. To get information on the quality of newly made wall, one must either examine thin layers of innermost wall (Fig. 2c) or physically constrain the young cell in a jacket (see Fig. 12b, below) so that it cannot expand in any direction, thereby keeping the microfibrils in place. After treatment in the jacket the appropriate subtractions yield the nature of the new material made during treatment. The passive reorientation of microfibrils from a transverse arrangement at the wall inner surface to an isotropic or longitudinal one on the outer surface (Fig. 2b) was described by Houwink and Roelofsen (1954), and constituted highly significant evidence that cell walls grew by the apposition of new mass at the inner surface ("multi-net growth"). This structure has been confirmed for *Nitella* (Green, 1960).

Evidence for a Cytoplasmic Framework

Independence of ordered synthesis from pre-existing wall was shown by growing groups of young cells (approximately 3 mm long, 6 plants per group) in constraining jackets in the presence of the drug colchicine (0.4%). These

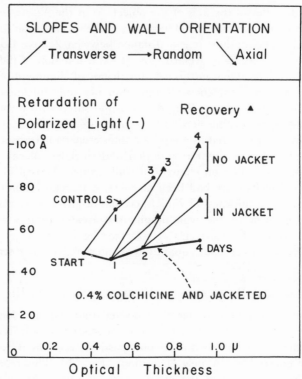

Fig. 11. An experiment wherein cells were treated with colchicine while in confining jackets to give random deposition of wall material (darkest line). It is seen that after 1, 2, and 4 days the treated cells were thicker but had not changed their action on polarized light (vertical axis) appreciably. This shows random wall deposition. Upon removal from the drug (recovery) the increase in thickness was accompanied by increased action on polarized light. This reveals an independence of the orientation of new synthesis from the previously deposited wall. Removal of the jacket during recovery improves the transverse order of the wall. Each point is the average of 6 cells.

treated cells increased their optical thickness or mass per area some 25% during the first day and an additional 25% the second day, while the action on polarized light remained essentially unchanged (Fig. 11). This showed the deposition of random material (Fig. 6) to a depth of 0.3 μ of optical thickness, or approximately 0.4 $\mu g/mm^2$. After 1 or 2 days some groups were allowed to recover by removing the drug or by removing the drug and the jacket. In all cases transverse order in the recovering walls improved during the following 2 days, revealing the return of transverse synthesis upon a random substratum. The transverse order was better in those cells recovering without the jacket, leading us to suspect that the deformations or strains of growth had an influence on ordered synthesis. These cells elongate, rather than get fat, presumably because they retain pretreatment transverse order. In control cells the over-all transverse order stays of the same character (same slope in Fig. 11) as wall thickness increases. Transverse order in the cells recovering without a jacket approaches that in control cells. It was concluded that there was a cytoplasmic framework responsible for transverse synthesis and that it operated independently of pre-existing wall texture. The proteinaceous nature of the framework is indicated by the fact that it is reversibly disturbed by colchicine, a drug which "dissolves" proteinaceous mitotic spindles (Eigsti and Dustin, 1955). Independence of oriented synthesis from the adjacent pre-existing wall texture is also indicated in the cross-fibrillar wall of *Chaetomorpha*, where a random lamella formed in a preswarming stage may be covered up by oriented lamellae if sporulation does not occur (Frei and Preston, 1961a, b).

The Cytoplasmic Framework—Its Action in Orienting Synthesis

The observed order in newly synthesized cell wall microfibrils is a function of the direction in which microfibrillar crystals were begun and the direction of their subsequent growth. Because the fibrils are crystalline they tend to be straight and one would expect their initial orientation to be of primary significance, although it is possible that both features of formation are controlled. Microfibrils are believed to grow at one or both of their ends because they are occasionally found in bent configurations difficult to reconcile with anything but tip growth (Preston, 1961a). The same conclusion had been reached by Colvin *et al.* (1957) for the extracellular growth of microfibrils of the bacterium *Acetobacter xylinum*, where short fibrils were shown to increase in length, but not diameter, with time. The intercalation of new glucose residues into cellulose chains is unlikely. The precise chemistry of cellulose formation is still under investigation. Hestrin (1961) states that in *Acetobacter* hexose phosphate is converted into cellulose chains by an enzyme known to be anchored to the cell. Microfibrils appeared only in the external medium and it was postulated that a diffusing form "probably a lone cellulose molecule" assembles spontaneously into the fibrous crystalline form.

A directing of crystal growth, although not of the orientation of the initial crystal axes, has been described by Glimcher (1959) for the mineralization of bone. The natural form of collagen (but not reconstituted collagen of different axial period) promotes the crystallization of hydroxyapatite in the center of the collagen fibril and sterically confines the growing crystal so that it ultimately takes a position with its long axis parallel to that of the collagen fibril. Thus a protein has been shown to influence the orientation of nonproteinaceous crystal growth; an action of this sort is postulated for the cytoplasmic framework with regard to the oriented synthesis of microfibrils. The framework would be expected to be located in the cytoplasm adjacent to the wall, peripheral to the chloroplasts which are stationary except for their own elongation and division. This region of the cell is not in the protoplasmic stream.

Nitella internode cells during their development increase in length many thousandfold and increase in diameter approximately tenfold. Because transverse synthesis is maintained throughout, the cytoplasmic framework involved must expand to cover new cell surface—both longitudinally and laterally—and must do so without becoming disorganized. A self-extending latticework, somewhat like that found in the cortical cytoplasm of protozoa (Lwoff, 1950; Tartar, 1961), might account for the essential constancy of transverse synthesis. It struck us, however, that the framework could hardly be so perfectly aimed at the time of internode cell formation and so perfect in its self-extension that it could function accurately throughout a ten thousandfold increase in cell surface. We assumed that some directed feature or features of cell growth would feed back on the framework to maintain its proper alignment, and we have therefore attempted to find out how, in effect, the cytoplasm recognizes the transverse direction.

Influences on the Orientation of the Cytoplasmic Framework

A small patch of cytoplasm could, in principle, base its directed synthesis on a number of oriented features of its immediate environment.

1. Synthesis could take place according to the direction of maximum curvature, although the radius of curvature at the synthetic site is enormous (up to 200 μ) compared to the dimensions of cytoplasmic structures that could possibly detect the direction of curvature.

2. It could take place along the lines of maximum stress in a cylinder wall under tension, the transverse direction (Fig. 2a). Stress is a force per area in contrast to strain, which is a deformation caused by stress. This mechanism has been considered unlikely on the grounds that the cytoplasm could not directly determine the direction of maximum stress, when this direction was different from the direction of maximum strain. In general, stresses are measured by evaluating *strains* in bodies constructed of physically isotropic materials. This procedure, common in industry (Teague and Blau, 1956), is unlikely to be followed by the cytoplasm, because the nearby strains in the wall are in highly anisotropic materials (Green and Chen, 1960).

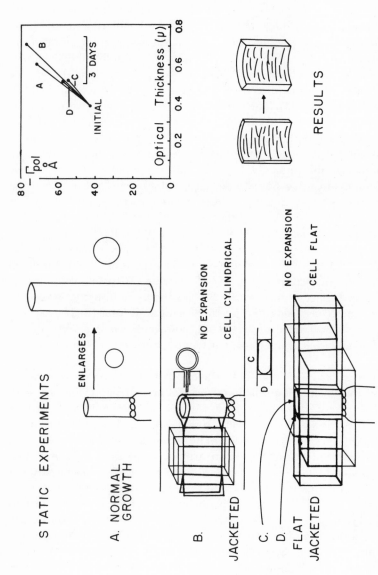

Fig. 12. A comparison of character of wall synthesized under various conditions. In all cases increase in thickness was accompanied by increased action on polarized light, revealing transverse synthesis. These experiments indicated that static geometry did not play an important role in ordering wall synthesis.

3. It could take place at some fixed angle to the direction of maximum strain (deformation) in the "stationary" cytoplasm adjacent to the wall. The *Nitella* cell twists as it grows and thus the direction of maximum strain is typically oblique. The direction of maximum strain, unlike the two above directions, could be readily determined by a patch of cytoplasm. In fact the chloroplast files reflect the direction of maximum strain. They run in steep helices up the cell (Fig. 13). The protoplasmic stream flows along the helical lines formed by the chloroplasts (Figs. 13 and 14), and thus a relation of oriented synthesis to either the chloroplast file direction or the streaming direction would be ultimately a relation to the direction of maximum strain.

4. Another possibility would be a synthesis parallel to the transverse strain. If the cytoplasmic framework contained strong elements that went around the cell as loops, then the increase in circumference of the cell would tend to make transverse any loops that were oblique (Fig. 29). This mechanism was proposed (Diehl *et al.*, 1939; Roelofsen, 1958, 1959) to account for transverse order in the wall in general. It would explain any improvement in transverse order in a wall growing through the deposition of poorly ordered loops of microfibrils. Because the degree of transverse order falls continuously from the inner surface of the wall in *Nitella* (due to multi-net growth), this "rectifying" mechanism does not operate in the wall itself (there is no improvement to be accounted for). It could, however, keep transverse a loop-containing cytoplasmic framework that tended to become oblique or poorly ordered.

We first attempted to find an influence of cell geometry on the orientation of synthesis by constraining cells in cylindrical or flattened cylindrical configuration. In the absence of expansion, transverse synthesis continued no matter whether the wall was normally round (see Fig. 12b), flat (Fig. 12c), or of increased curvature (Fig. 12d). The new material deposited was all of the same order of transverse texture (similar slope in the graph in Fig. 12), and similar to the over-all texture of control walls growing without constraint. This led us to conclude that features of static geometry did not play an important role in orienting synthesis. The recovery of order in synthesis in jacketed cells after colchicine-jacket treatment was thus considered to be "spontaneous" rather than directly based upon the curvature of the cell. The much better recovery in dejacketed cells was presumably based on some feature of cell expansion other than cell geometry.

Role of Strain in the Orientation of the Cytoplasmic Framework

Because the recovery experiments suggested a role for strain and because the oriented chloroplast files reveal an action of strain on the cytoplasm ad-

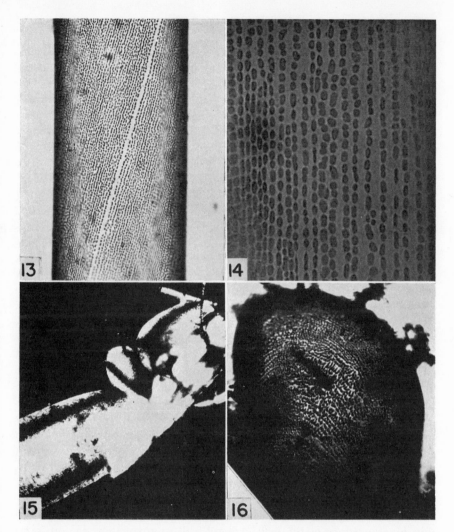

Fig. 13. A living internodal cell showing the oblique chloroplast files. The light line running up and to the right is the striation, a gap in the monolayer of plastids. Such lines separate the ascending and descending streams of protoplasm. These lines can be detected in single wall thicknesses when seen in polarized light (Figs. 22, 23, and 24). This obliquity of chloroplast files is typical. 100×.

Fig. 14. Chloroplast files under higher power. The obliquity is not shown. Note inter-digitating files (*top*, *center*) and the fact that the plastids elongate in the file axis. 500×.

Fig. 15. A living cell seen in polarized light (crossed prisms). The main cell runs from lower left to upper right. In the center is a fingerlike outgrowth, at upper right a flaplike outgrowth. Orientation of the wall microfibrils (high index of refraction) is perpendicular to the double dark shadows in the cell. The wall has become reoriented in the fingerlike outgrowth but not in the flaplike outgrowth. The birefringent jacket (dialysis tubing) obscures details on the lower part of the cell. 40×.

Fig. 16. A live small induced lateral. Note that in the center the chloroplasts are elongating along the new axis of growth (maximum strain). The original chloroplast files can be seen as semicircles. 250×.

jacent to the wall, we investigated the orientation of newly made wall (Fig. 2c) under conditions where the adjacent cytoplasm had different strain environments.

The production of lateral fingerlike outgrowths with wall texture transverse to the new axis of extension supported the view that microfibrils were put down perpendicular to the direction of maximum strain (deformation); see Figs. 9, 10, 15, 17a. When cells were partially ligated and directly stretched by a pulley and weights so that all the strain was longitudinal (increase in circumference was inhibited), as shown in Fig. 17b, transverse order could be readily seen in the inner part of the cell wall (Figs. 20, 21), indicating synthesis normal to the strain. This procedure incidentally enhanced passive reorientation of fibrils in certain outer parts of the wall so total wall retardation was low or even reversed (longitudinal) in sign (Fig. 18). The weights applied to stretch the cell were such as to rotate the direction of maximum stress in the wall from the transverse direction to the longitudinal direction (Green and Chen, 1960), weighing against possibility number 2 (transverse stress), already considered unlikely on theoretical grounds.

The above observations support the view that synthesis takes place approximately at right angles to the direction of maximum strain, and if the cytoplasm adjacent to the wall contained long elements that were flow-oriented longitudinally by the distortions of elongation itself (not the protoplasmic stream), these long elements would synthesize microfibrils perpendicular to their long axis. This would be a positive feedback mechanism.

A third observation supporting the view that newly made microfibrils were put down perpendicular to the direction of maximum strain came from cells permitted to grow in dilute (0.25%) solutions of colchicine without being jacketed. The cells grew fat and were very twisted, as in Fig. 4 (left), and examination of the walls revealed that the inner portion showed a highly oblique extinction. In a cell so twisted that the chloroplast files were 35° off-axial, the single wall thickness extinguished some 20° off-transverse and thin inner portions of the wall extinguished approximately 35° off-transverse, that is, normal to the direction of the chloroplast files (Figs. 17c, 24). Here the strains leading to obliquity of the peripheral cytoplasm (chloroplast files) had an equivalent effect on the obliquity of the synthetic machinery for microfibrils. Less extreme oblique extinctions of this type (for single thicknesses rather than inner parts of wall) are found in untreated *Nitella opaca* (Probine and Preston, 1958).

It is surprising that the above experiments suggesting an overriding role for the direction of maximum strain are in part confounded by observations on growing normal cells. In normal cells the synthesis is more nearly transverse than one would predict if synthesis were to be normal to maximum

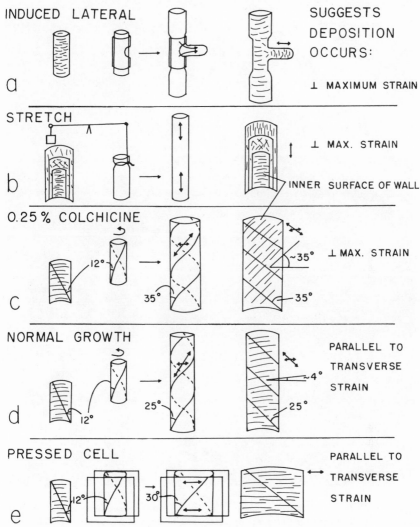

FIG. 17. Experiments presenting the peripheral cytoplasm with a variety of strain environments. The initial state of the wall and the nature of the experiment is given at left. The structure of the wall after treatment is given at right as is the simplest explanation to account for the orientation of deposition in each case. The crossed arrows show the strain pattern.

strain as revealed by the chloroplast files. The obliquity of extinction should vary directly as the obliquity of chloroplast files. This variation is found to some extent (it is not a fixed relation) in *N. opaca* (Probine and Preston, 1958) but not in *N. axillaris*. As shown in Figs. 17d and 22, a normal cell

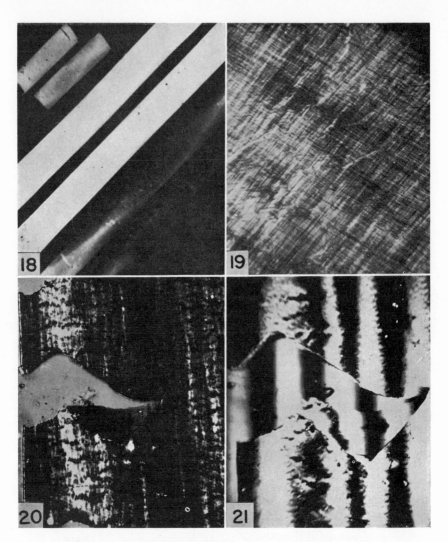

Fig. 18. *Upper left*, two control cells. In the center, two controls after growth. At *lower right*, two walls from cells that had been stretched to prevent increase in diameter (Fig. 17b). The low brightness of the stretched walls resulted from accelerated multi-net growth. That synthesis remained transverse is revealed in Figs. 20 and 21. Reorientation is enhanced in zones away from the striation. 12×.

Fig. 19. The crossed-fibrillar wall of *Valonia* seen in polarized light. 160×.

Fig. 20. Wall from Fig. 18 seen in polarized light, compensator rotated. Areas darker than the field have transverse orientation, areas lighter have longitudinal orientation. The dark triangle at center shows transverse order although it represents only about ⅕ the total wall thickness (Fig. 21). It is the most recently built wall. 100×.

Fig. 21. Same wall as in Fig. 20, in the interference microscope. Wall thickness is proportional to the displacement of fringes to the right. The triangular piece of Fig. 20 is shown to be a thin inner part of the wall and of good transverse texture. 130×.

222

with the chloroplast files some 26° off-axis has the extinction position at most 4° off-transverse, and typically the cells extinguish so to show essentially transverse orientation (Green, 1959) despite wide variation (5–20°) in degree of twist. This suggested that the normal cell has a rectifying mechanism which is overridden by the strains of induced lateral formation and of stretched cells. In colchicine it is presumably broken down by the drug—so that in these cases orientation is normal to maximum strain (Fig. 29).

Nature of the Rectifying Mechanism

With a role for curvature not supported by previous tests, the rectifying device could be based on transverse strain in the manner of the "loop-rectifying" device proposed by Roelofsen and others to operate in the wall itself to improve transverse order of microfibrils. An oblique loop or ring around a cylinder will be brought into the transverse position as the cylinder increases in diameter (Fig. 29). Evidence supporting this idea came from experiments where cells grew in circumference only as they were pressed between glass plates (Fig. 17e). An average increase in circumference of 70% was obtained with some instances of a 200% increase. As shown in Fig. 23, extinction of single layers was transverse and the degree of transverse order in the wall was extremely high throughout. In Fig. 25 the retardation of a thin innermost layer is 30 Å. The thickness of this layer is revealed in Fig. 26, where the displacement of fringes in the layer is only 0.1 wavelength. A corresponding innermost area from a jacketed cell (no expansion at all) shows that the district compensating at 30 Å (Fig. 27) is considerably thicker (0.18 wavelength) (Fig. 28). Despite the pure transverse strain (evidenced by longitudinal cracks as shown in Fig. 23), the synthesis was highly transverse—exactly in the direction of maximum strain. This confounds the first three experiments and forces the view that there is a hierarchy of mechanisms—the loop-rectifying mechanism operating in normal growth and in pressed cells (Fig. 29) and this mechanism being overridden by severe strains in stretched cells and induced laterals, the mechanism also being disabled by colchicine.

It is possible to give a criterion for severity of strain.

Moderate strain affects the chloroplasts differently from what may be called drastic strain. Both pressed-cell growth (Figs. 17e, 24) and flaplike outgrowths (Fig. 15, upper right) display moderate strain because the chloroplast files retain their identity and are reoriented toward the new direction of maximum strain by rotating *en masse*, as shown by the change in pitch of the diagonal line in Fig. 17e. The files never come to lie in the new direction of maximum strain (transverse) but only approach this direction. The tilting is remarkable in that it illustrates a slipping of the rotating chloroplast layer

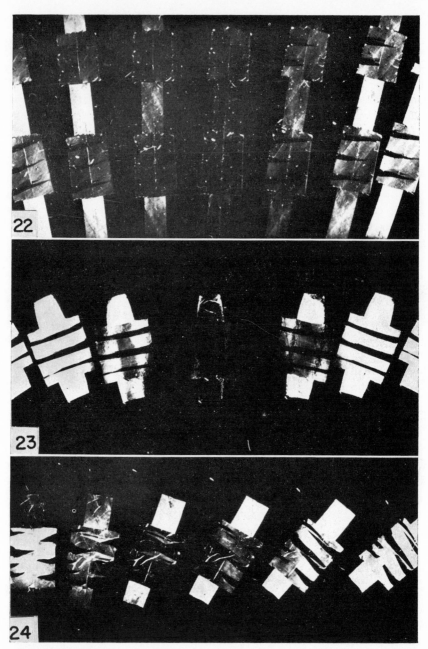

Figs. 22–24. Multiple exposures of rotated cell walls taken in polarized light (prisms crossed, transmission planes parallel to the edges of the page). The film was advanced between exposures. The narrow parts of the walls are collapsed cylindrical walls (double thickness) while the broad regions are single thicknesses. Horizontal tears in the wall have been used at higher power to determine the optical character of the most recently synthesized portion (Fig. 2c).

Fig. 22. A normal wall. Note that the brightness of double layers is symmetrical around

against the cell wall whose rotation is prevented. Rotation occurs in the free-growing edges of the pressed cell. The appearance of the chloroplast files themselves is normal (Fig. 14) and many new files are produced by interdigitation of elongating files to keep the new area covered with a typical monolayer. Lateral extension of this type does not drastically change the chloroplast pattern. The pattern of wall synthesis likewise is not much changed as the transverse texture is continued or even improved (Figs. 10, 25, 26), despite the change in the direction of maximum strain (Fig. 29).

Severe strain is found in fingerlike outgrowths where the pattern of wall synthesis is altered so as to be perpendicular to the new axis of extension. The effect on the chloroplasts also is striking in that individual chloroplasts begin to elongate immediately in the direction of maximum strain (see Fig. 16), The old files can still be recognized as semicircular extensions into the lateral but chloroplast elongation no longer follows the file axis. The difference between fingerlike and flaplike outgrowths, which can occur together in the same cell (Fig. 15), appears to be based on the severity of strain. In both cases the direction of maximum strain is lateral. If the strain is modest, wall synthesis continues as before although now it is in the direction of maximum strain. If the strain is extreme the direction of wall synthesis becomes normal to the direction of maximum strain. Preparations of this type present a dilemma which, to date, we have been able to explain only through a hierarchy of responses based on transverse and maximum strain.

A Provisional Model

One can assume that the cytoplasmic framework consists of long elements which start or guide microfibrillar formation normal to the element's long axis (Fig. 29). Under conditions of normal growth or pressed-cell growth

---

the center exposure. The single thicknesses are symmetrical around a point between the center figure and the one to the left of it, revealing a slight (3°) obliquity of extinction. The degree of twist of the wall is seen in the oblique lines in the extreme left and right figures and is about 26°. Inner wall surface down; 5° turn between exposures. 12×.

FIG. 23. A wall from a pressed cell that had increased in circumference 200% at constant length. Note symmetrical brightness of the center region of the single thickness, the free-growing part of the wall. Lack of symmetry for other regions is due to curvature of the wall. Inner wall surface up; 10° turn between exposures. 9×.

FIG. 24. A wall from a cell grown in colchicine, no jacket. The difference in extinction behavior between single and double thickness is so great that only one side of the rotation is shown. Note that double thicknesses extinguish on the cell axis, while single layers extinguish 20° away, and thinner inner parts of the wall—white areas in the single thickness in the third figure from left—extinguish some 35–40° off the cell axis. This shows synthesis at right angles to the obliquity of the strain occurring in cytoplasm—as revealed in the obliquity of chloroplast files. The lines running through the single thicknesses are the striations in the chloroplast pattern and are 35° off axis. 9×.

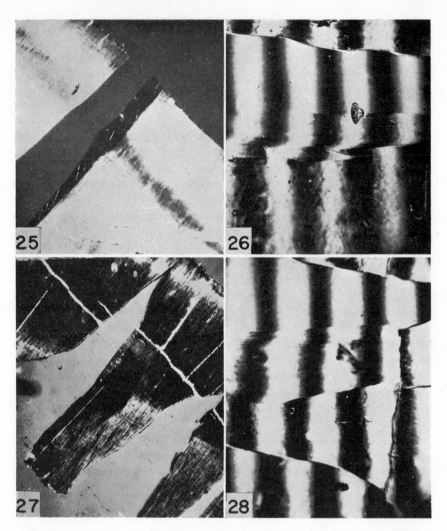

Fig. 25. The black portion is a thin inner layer of a pressed-cell wall being compensated at 30 Å. 100×.

Fig. 26. Same wall as in Fig. 25, but rotated 45° to the right and enlarged. The thickness of the specimen is reflected in the displacement of the vertical interference fringes to the left. The displacement here is about 0.1 wavelength for the thin layer, about 0.6 for the entire wall. 200×.

Fig. 27. As in Fig. 25, but from a comparable wall which increased in wall thickness while confined in a jacket. The dark central portion is compensated at 30 Å. 100×.

Fig. 28. Same wall as in Fig. 27, rotated 45° to the right and enlarged. The region compensating at 30 Å in Fig. 27 is seen to displace the fringes about 0.18 wavelength. Because equivalent action on polarized light requires more thickness than in the pressed-cell case, transverse order here is poorer than in the pressed-cell case. This shows that in pressed cells growth with purely transverse strain does not lower the transverse order but actually appears to improve it. 170×.

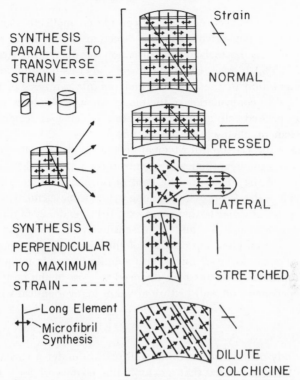

Fig. 29. A model designed to account for the fact that under some conditions (including normal growth) wall synthesis stays essentially transverse despite considerable variation in the direction of maximum strain, while under other conditions synthesis is at right angles to the direction of maximum strain. The latter behavior suggests a coupling of the synthesis to long flow-oriented bodies. The former behavior suggests a coupling of the synthesis to bodies kept in a fixed orientation by the tautness of transverse bands in the cytoplasm. Under the conditions in the top two types of growth, the effect of the transverse bands predominates and synthesis is transverse. In the lower three cases, the lateral bonding is assumed ineffective and wall synthesis is normal to the direction of maximum strain.

these elements are assumed kept in the longitudinal direction by a loop-rectifying device based on transverse strain. The elements are assumed to have bridges connecting them to give the cytoplasm looplike character. The loops are kept taut by transverse strain (increase in circumference). Under these conditions wall synthesis is transverse to the cylindrical axis and is independent of the direction of maximum strain which is variably oblique.

Under conditions of severe strain (fingerlike laterals and stretched cells) the cross-bonding between elements is considered ineffective and the long elements are then flow-oriented to the direction of maximum strain. This results in synthesis normal to the direction of maximum strain. Dilute colchicine (0.25%) is assumed to inactivate the cross-bonding also, permitting flow-orientation of the elements and oblique synthesis of microfibrils (Fig. 29). Further confirmation of the model, such as obtaining longitudinal deposition in pressed cells growing in dilute colchicine, is deemed necessary before the elements themselves are searched for.

Dr. Inoué of the Dartmouth Medical School has drawn my attention to a parallel between the hypothetical elements in the model and phragmoplasts. Three aspects of the model for the control of microfibrillar orientation in elongating side walls, (a) presence of long cytoplasmic elements which make microfibrils perpendicular to their long axis, (b) a tendency of these elements to become bonded laterally, and (c) a sensitivity of this alignment to colchicine, are all well-known properties of the fibers of phragmoplasts. The latter fibers appear to be identical with spindle fibers. It is thus suggested that the cytoplasmic elements here deduced to exist in the peripheral cytoplasm by experiments on wall texture are akin to, or identical with, spindle-fiber protein.

Discussion

From a study of the orientation of wall texture under a variety of conditions, it has been concluded that a cytoplasmic framework for wall synthesis exists in *Nitella*. This framework appears to be influenced in its orientation not by pre-existing wall texture but by strain. That this framework is not identical with the visible chloroplast and protoplasmic streaming pattern (also influenced by strain) is indicated by the observation that in normal growth or pressed-cell growth the wall texture remains nearly transverse despite pitch variations of more than 20° in the chloroplast files and stream. The framework thus presumably exists in a still more peripheral part of cytoplasm, the region between the chloroplasts and the wall. Direct cytoplasmic control over the orientation of cellulose has long been assumed in walls with a crossed-fibrillar texture, because shifts in microfibrillar orientation are sharp and frequent (see Fig. 19). Preston (1961a) has recently discussed the appearance of *oriented* proteinaceous strands *prior* to the deposition of microfibrils in that orientation. These effects are seen in *Chaetomorpha* (Nicolai and Preston, 1959) and *Valonia* (Preston, 1961).

That pre-existing wall has at least some influence on the cytoplasm in crossed-fibrillar forms, in apparent contrast with *Nitella*, is indicated in the observation that in plasmolyzed cells of *Chaetomorpha* the free part of the cytoplasmic surface makes microfibrils at random, while upon deplasmolysis and re-establishment of contact with the original wall typically ordered lamellae of microfibrils are produced (Frei and Preston, 1961a).

The cytoplasmic component most likely to be involved in control of wall synthesis is the endoplasmic reticulum (ER). Porter and Machado (1960) and Porter (1961) have provided electron micrographs of sections of plant cells, which show the ER in very close association with the newly formed cross-wall after cell division, the spiral thickening of protoxylem, and the epidermal cell wall. Small bits of the ER are believed to be "externalized" into the wall during wall synthesis by apposition—a process suggested by Myers *et al.* (1956). While the close association between the wall and the ER is evident in cross sections of cells, a tangential section grazing the wall and adjacent cytoplasm in an "onion root tip cell" (Porter, 1961) reveals the ER in an apparently random network. Presumably the cell in question was carrying out ordered synthesis of wall of either the primary or secondary type. This fortuitous test of the relation between the geometrical organization of the ER and adjacent wall texture reveals no correspondence, and, in the absence of other information, it is concluded that the cytoplasmic framework consists of elements smaller than the recognizable features of the ER.

A cytoplasmic framework in the desmid *Micrasterias* (Green Algae) has been studied by the Finnish workers Kallio and Waris. The cell consists of two semi-cells located symmetrically on either side of the nucleus-containing isthmus. At cell division, after mitosis, the isthmus makes two new semi-cells, one for each nucleus. Thus each daughter cell has one parental semi-cell and one new semi-cell. When enucleate parental semi-cells are made (by centrifugation) a new semi-cell is nonetheless produced, although of greatly simplified geometry. It was concluded that three cytoplasmic axes extend through the isthmus from the parental semi-cells and serve as guides for the construction of the new semi-cells. A "defect mutation" (Kallio, 1951), where the semi-cells each lack the same lobe, is believed to reflect the loss of one cytoplasmic axis. Back mutation involves the doubling of one of the axes. That the mutation is cytoplasmic rather than nuclear is supported by the observation that it can be induced in high yield by centrifugation and that two nuclei are involved in divisions leading to the appearance of the mutation (Kallio, 1951). Some visible evidence for the three axes is presented by Waris (1950, 1951).

The most striking cytoplasmic frameworks occur in the cortical cytoplasm of ciliates. In *Stentor* Tartar has shown that disorganization of the network into patches yields animals with a corresponding lumpy form. Gradual rearrangement of the patches reconstitutes the animal. "The main point is that intrinsic polarity persists in the striped ectoplasm, no matter how oriented" (Tartar, 1961, p. 196). Variation in the striped ectoplasm is associated with all the prominent features of morphogenesis in the animal both in normal division and regeneration. It has been shown in *Paramecium* that aberrations in the ectoplasmic network can persist through many cell divisions, revealing a semi-autonomous morphogenetic behavior for the cortex (Sonneborn and Dippel, 1960). The non-nuclear nature of the aberration is shown by the persistence of the unusual growth pattern despite the introduction of a wild-type nucleus for the original one.

The elucidation of ciliate morphogenesis in terms of the behavior of visibly self-duplicating kinetosomes unfortunately falls short of an ultimate explanation of development in the organisms. The impressive activities of these bodies—oriented growth and division, differentiation into trichocysts, local oral structure formation, disappearance, etc.—are in turn controlled by entirely mysterious properties of the underlying cytoplasm (Lwoff, 1950).

Evidence for cytoplasmic frameworks is widespread enough, and the observed properties of some of them are complex enough, to suggest that the physical texture of cortical cytoplasm is an oriented feature that could be compounded in the development of multicellular organisms to account for relatively gross morphogenetic phenomena.

Summary

In plants the elongation of the organ can be directly related to the elongation of the single cell. Elongation of the *Nitella* cell, which occurs throughout its length, is based on a physical anisotropy of the wall. The transverse arrangement of strong cellulose microfibrils renders the wall strong in the transverse direction, so that it grows primarily in length. Elongation involves the continued synthesis of microfibrils in the transverse direction. The orientation of synthesis of new fibrils can be shown to be independent of the arrangement of the pre-existing fibrils and thus oriented synthesis is based on order in the synthetic machinery or a "cytoplasmic framework."

The cytoplasmic framework was relatively little influenced by such static features of cell geometry as curvature, but did show variation when the strain (deformation) pattern of growth was altered by experimentation. There were two types of response to strain. In three different experiments synthesis took place at right angles to the direction of maximum strain, indicating a coupling of synthesis to hypothetical strain- or flow-oriented long elements in the cortical cytoplasm. In two other cases, including normal growth, synthesis was not normal to the direction of maximum strain but was in the direction of transverse strain. This latter behavior suggested that the long elements were, in these cases, kept at a fixed angle to the cell axis by a mechanism wherein transverse bands in the cytoplasm (presumably formed by cross-linking the long elements) were kept taut by the increase in circumference (transverse strain) of the cell. The long elements have many properties of phragmoplast (spindle) fibers. The strain-sensitive behavior of the synthetic machinery of the peripheral cytoplasm appears to be one mechanism whereby cellular morphogenesis can be maintained or altered.

Acknowledgments

The work reported here was supported by a grant from the National Science Foundation (G-12919). Mrs. Mary Eager contributed greatly to the success of the original experiments.

References

Bayley, S. T., and Setterfield, G. (1957). The influence of manniotol and auxin on growth of cell walls in *Avena* coleoptiles. *Ann. Botany (London)* [N.S.] 21, 633–641.

Bennett, H. S. (1950). The microscopical investigation of biological materials with polarized light. *In* "Handbook of Microscopical Technique" (R. M. Jones, ed.), pp. 591–677. Harper (Hoeber), New York.

Berrill, N. J. (1949). The polymorphic transformations of *Obelia*. *Quart. J. Microscop. Sci.* 90, 235–264.

Berrill, N. J. (1961). "Growth, Development, and Pattern." Freeman, San Francisco, California.

Castle, E. S. (1955). The mode of growth of epidermal cells on the *Avena* coleoptile. *Proc. Natl. Acad. Sci. U. S.* 41, 197–199.

Colvin, J. R., Bayley, S. T., and Beer, M. (1957). The growth of cellulose microfibrils from *Ocetobacter xylinum*. *Biochim. et Biophys. Acta* 23, 652–653.

Das, N. K., Patau, K., and Skoog, F. (1956). Initiation of mitosis and cell division by kinetin and indole acetic acid in excised tobacco pith tissue. *Physiol. Plantarum* 9, 640–651.

Delay, C. (1957). Action de la colchicine sur la croissance et la différenciation de l'appareil végétatif de *Chara vulgaris* L. *Compt. rend. acad. sci.* 244, 485–487.

Diehl, J. M., Gorter, C. J., van Iterson, G., and Kleinhoonte, A. (1939). The influence of growth hormone on hypocotyls of *Helianthus* and the structure of the cell walls. *Rec. trav. botan. neerl.* 36, 709–798.

Dodds, G. S. (1930). Row formation and other types of arrangement of cartilage cells in endochondrial ossification. *Anat. Record* 46, 385–399.

Eigsti, O. J., and Dustin, P., Jr. (1955). "Colchicine." Iowa State College Press, Ames, Iowa.

Erickson, R. O., and Sax, K. (1956a). Elemental growth rate of the primary root *Zea mays*. *Proc. Am. Phil. Soc.* 100, 487–498.

Erickson, R. O., and Sax, K. (1956b). Rate of cell division and cell elongation in the growth of the primary root of *Zea mays*. *Proc. Am. Phil. Soc.* 100, 499–514.

Foard, D. E., and Haber, A. H. (1961). Anatomic studies of gamma-irradiated wheat growing without cell division. *Am. J. Botany* 48, 438–446.

Frei, E., and Preston, R. D. (1961a). Cell wall organization and wall growth in the filamentous green algae *Cladophora* and *Chaetomorpha*. I. The basic structure and its formation. *Proc. Roy. Soc. (London)* B154, 70–94.

Frei, E., and Preston, R. D. (1961b). Cell wall organization and wall growth in the filamentous green algae *Cladophora* and *Chaetomorpha*. II. Spiral structure and spiral growth. *Proc. Roy. Soc. (London)* **B155**, 55–77.

FREY-WYSSLING, A. (1959). "Die pflanzliche Zellwand." Springer, Berlin.

GLIMCHER, M. (1959). Molecular biology of mineralized tissue with special reference to bone. *In* "Biophysical Science—A Study Program" (J. L. Oncley, *et al.*, eds.), pp. 359–393. Wiley, New York.

GREEN, P. B. (1954). The spiral growth pattern of the cell wall in *Nitella axillaris*. *Am. J. Botany* **41**, 403–409.

GREEN, P. B. (1958a). Concerning the site of the addition of new wall substances to the elongating *Nitella* cell wall. *Am. J. Botany* **45**, 111–116.

GREEN, P. B. (1958b). Structural characteristics of developing *Nitella* internodal cell walls. *J. Biophys. Biochem. Cytol.* **4**, 505–516.

GREEN, P. B. (1959). Wall structure and helical growth in *Nitella*. *Biochim. et Biophys. Acta* **36**, 536–538.

GREEN, P. B. (1960). Multinet growth in the cell wall of *Nitella*. *J. Biophys. Biochem. Cytol.* **7**, 289–296.

GREEN, P. B. (1962). Cell expansion. *In* "Physiology and Biochemistry of Algae" (R. A. Lewin, ed.), pp. 625–632. Academic Press, New York.

GREEN, P. B., AND CHEN, J. (1960). Concerning the role of wall stresses in the elongation of the *Nitella* cell. *Z. wiss. Mikroskop.* **64**, 482–488.

GUILLARD, R. (1962). Salt and osmotic balance. *In* "Physiology and Biochemistry of Algae" (R. A. Lewin, ed.), pp. 529–540. Academic Press, New York.

HABER, A. H. (1962). Nonessentiality of concurrent cell divisions for degree of polarization of leaf growth. I. *Am. J. Botany* **49**, 583–589.

HALE, A. J. (1960). The interference microscope as a cell balance. *In* "New Approaches in Cell Biology" (P. M. B. Walker, ed.), pp. 173–186. Academic Press, New York.

HEIMSCH, C. (1960). A new aspect of cortical development in roots. *Am. J. Botany* **47**, 195–201.

HESTRIN, S. (1961). The growth of saccharide macromolecules. *In* "Biological Structure and Function" (T. W. Goodwin and O. Lindberg, eds.), Vol. 1, pp. 315–323. Academic Press, New York.

HEYN, A. N. J. (1940). The physiology of cell elongation. *Botan. Rev.* **6**, 515–574.

HOLTFRETER, J. (1939). Studien zur Ermittlung der Gestaltungsfaktonen in der Organentwicklung der Amphibien. I und II. *Wilhelm Roux Arch. Entwicklungsmech. Organ.* **139**, 110–190; 227–273.

HOUWINK, A. L., AND ROELOFSEN, P. A. (1954). Fibrillar architecture of growing plant cell walls. *Acta Botan. Neerl.* **3**, 385–395.

JENSEN, W. A. (1961). Relation of primary cell wall formation to cell development in plants. *In* "Synthesis of Molecular and Cellular Structure," Growth Symposium No. 19 (D. Rudnïck, ed.), pp. 89–110. Ronald Press, New York.

KALLIO, P. (1951). The significance of nuclear quantity in the genus *Micrasterias*. *Ann. botan. soc. zool.-botan. fennicae Vanamo* **24**, 1–122.

KITCHIN, I. C. (1949). The effect of notochordectomy in *Amblystoma mexicanum*. *J Exptl. Zool.* **112**, 393–416.

KÜSTER, E. (1956). "Die Pflanzenzelle." Fischer, Jena.

LWOFF, A. (1950). "Problems of Morphogenesis in Ciliates." Wiley, New York.

MILLER, J. H. (1961). The effect of auxin and guanine on cell expansion and cell division in the gametophyte of the fern *Onoclea sensibilis*. *Am. J. Botany* **48**, 816–819.

MOOKERJEE, S. (1953). Experimental study of the development of the notochordal sheath. *J. Embryol. Exptl. Morphol.* **1**, 411–416.

MOOKERJEE, S., DEUCHAR, E. M., AND WADDINGTON, C. H. (1953). Morphogenesis of the notochord in amphibia. *J. Embryol. Exptl. Morphol.* **1**, 399–409.

MÜHLETHALER, K. (1961). Plant cell walls. *In* "The Cell" (J. Brachet and A. E. Mirsky, eds.), Vol. II, pp. 85–134. Academic Press, New York.

MYERS, A., PRESTON, R. D., AND RIPLEY, G. W. (1956). Fine structure in the red algae. I. X-ray and electronmicroscope investigation of *Griffithsia flosculosa*. *Proc. Roy. Soc. (London)* **B144**, 450–459.

NICOLAI, E., AND PRESTON, R. D. (1959). Cell-wall studies in the Chlorophyceae III. Differences in structure and development in the Cladophoraceae. *Proc. Roy. Soc. (London)* **B151**, 244–255

NORTHCOTE, D. H. (1958). The cell walls of higher plants: their composition, structure, and growth. *Biol. Revs. Cambridge Phil. Soc.* **33**, 53–102.

PORTER, K. R. (1961). The endoplasmic reticulum: Some current interpretations of its form and functions. *In* "Biological Structure and Function" (T. W. Goodwin and O. Lindberg, eds.), Vol. 1, pp. 137–154. Academic Press, New York.

PORTER, K. R., AND MACHADO, R. (1960). Studies on the endoplasmic reticulum. IV. Its form and distribution during mitosis in cells of onion root tip. *J. Biophys. Biochem. Cytol.* **7**, 167–180.

PRESTON, R. D. (1952). "The Molecular Architecture of Plant Cell Walls." Wiley, New York.

PRESTON, R. D. (1961a). Cellulose-protein complexes in plant cell walls. *In* "Macromolecular Complexes," 6th Ann. Symposium Soc. Gen. Physiologists (M. V. Edds, ed.), pp. 229–253. Ronald Press, New York.

PRESTON, R. D. (1961b). *In* "Recent Advances in Botany," 9th Intern. Botan. Congr., pp. 234–238. Univ. of Toronto Press, Toronto.

PROBINE, M. C., AND PRESTON, R. D. (1958). Protoplasmic streaming and wall structure in *Nitella*. *Nature* **182**, 1657–1658.

PROBINE, M. C., AND PRESTON, R. D. (1961). Cell growth and the structure and mechanical properties of the wall in internodal cells of *Nitella opaca*. I. Wall structure and growth. *J. Exptl. Botany* **12**, 261–282.

PROBINE, M. C., AND PRESTON, R. D. (1962). Cell growth and the structure and mechanical properties of the wall in internodal cells of *Nitella opaca*. II. Mechanical properties of the walls. *J. Exptl. Botany* **13**, 111–127.

RAY, P. M. (1961). Hormonal regulation of plant cell growth. *In* "Control Mechanisms in Cellular Processes," 7th Ann. Symposium Soc. Gen. Physiologists (D. M. Bonner, ed.), pp. 185–212. Ronald Press, New York.

ROELOFSEN, P. A. (1958). Cell-wall structure as related to surface growth. *Acta Botan. Neerl.* **7**, 77–89.

ROELOFSEN, P. (1959). "The Plant Cell Wall." Borntraeger, Berlin.

SCHMITT, F. O. (1959). Interaction properties of elongate protein macromolecules with particular reference to collagen (tropocollagen). *In* "Biophysical Science—A Study Program" (J. L. Oncley, ed.), pp. 349–358. Wiley, New York.

SETTERFIELD, G., AND BAYLEY, S. T. (1959). Deposition of cell walls in oat coleoptiles. *Can. J. Botany* **37**, 861–870.

SETTERFIELD, G., AND BAYLEY, S. T. (1961). Structure and physiology of cell walls. *Ann. Rev. Physiol.* **12**, 35–62.

SLAYTER, G. (1962). Two-phase materials. *Sci. American* **206**, 124–134.

SONNEBORN, T., AND DIPPELL, R. V. (1960). The genetic basis of the difference between single and double *Paramecium aurelia*. *J. Protozool.* **7**, Suppl., 26.

STEWARD, F. C., AND MÜHLETHALER, K. (1953). The structure and development of the cell-wall in the Valoniaceae as revealed by the electron microscope. *Ann. Botany (London)* [N.S.] **27**, 295–316.

TARTAR, V. (1961). "The Biology of Stentor." Pergamon Press, New York.

TEAGUE, J. M., JR., AND BLAU, H. H. (1956). Investigations of stresses in glass bottles under internal hydrostatic pressure. *J. Am. Ceram. Soc.* **39**, 229–238.

WARDROP, A. B. (1956). The nature of surface growth in plant cells. *Australian J. Botany* **4**, 193–199.

WARIS, H. (1950). Cytophysiological studies on Micrasterias. II. The cytoplasmic framework and its mutation. *Physiol. Plantarum* **3**, 236–246.

WARIS, H. (1951). Cytophysiological studies on Micrasterias. III. Factors influencing the development of enucleate cells. *Physiol. Plantarum* **4**, 387–409.

WEISS, P. (1955). Nervous system. *In* "Analysis of Development" (B. H. Willier, P. A. Weiss, and V. Hamburger, eds.), pp. 346–401. Saunders, Philadelphia, Pennsylvania.

WEISS, P. (1959). Cellular dynamics. *In* "Biophysical Science—A Study Program" (J. L. Oncley, ed.), pp. 11–20. Wiley, New York.

WEISS, P. (1962). From cell to molecule. *In* "The Molecular Control of Cellular Activity" (J. M. Allen, ed.), pp. 1–72. McGraw-Hill, New York.

WILSON, E. B. (1889). The embryology of the earthworm. *J. Morphol.* **3**, 387–462.

WILSON, K. (1957). Extension growth in primary cell walls with special reference to *Elodea canadensis*. *Ann. Botany (London)* [N.S.] **21**, 1–11.

Tissue Interaction and Specific Metabolic Responses: Chondrogenic Induction and Differentiation

JAMES W. LASH

Department of Anatomy, School of Medicine, University of Pennsylvania, Philadelphia, Pennsylvania

The quest for chemical manifestations of embryonic induction has consumed the time and energies of many biologists. From Spemann and Mangold's original observations on embryonic induction (1924) to the present emphasis on molecular interactions, however, there has yet to come a completely substantiated instance of true biochemical induction that has developmental significance. The notion of a molecular intercession in the causation of morphological events has preoccupied many embryologists to the point of obscuring basic problems of embryonic induction.

The term "embryonic induction" has been defined by many researchers, but each definition relies for its validity upon its operational merits. That is to say, biological definitions are most satisfactory when they possess a certain degree of operational feasibility. Definitions that do not lend themselves to experimental testing in a critical manner soon cease to be valid since they do not yield satisfactory explanations. To "induce" means one thing to an embryologist channeled toward the earliest events of development, e.g., the formation of neural-like tissue from a relatively nondescript gastrula-derived ectoderm. It means yet another thing to one involved in the rather restricted area of postgastrulation morphogenesis or organogenesis. Currently, with the explosive advances in the mechanisms of metabolic expression in microorganisms, the term "induction" acquires a new and more specific meaning.

According to the concepts advanced by Jacob and Monod (1961), a particular gene complex, or operon, can determine the activity of well-prescribed and well-known chemical reactions in microorganisms by imparting genetic information to cytoplasmic biosynthetic machinery. These reactions result in the formation and ascendance of molecular patterns, which in turn determine the metabolic events the microorganism can or cannot perform. To the various concerned fields of biology, it would appear that induction now acquires newly defined and operationally feasible proportions. The specific meaning of induction now indicates the direction, or the control of the expression, of genetic information. Qualifications of repression, derepression, and feedback inhibition (Davis, 1961), only add to the validity of this concept as a quantitative operational aspect of cellular activity. The problem faced by the embryologist is that he is not yet ready to approach the problems of induction and differentiation on such a sophisticated level. If concepts such as this would have any meaning for embryologists with their current lack of adequate genetic material, then it should be paraphrased in a less specific manner to say that embryonic induction is the interaction between two tissues that results in the acquisition of specific metabolic patterns. This definition is not new, and is essentially the same as that given by Spemann (1938) and other embryologists more recently (Weiss, 1947; Needham, 1950; Grobstein, 1954). The metabolic aspect of induction has been expressed previously by many investigators in microbiology (Cohen, 1954). The inductive event itself is yet obscure, but the metabolic patterns created (i.e., differentiation) are better known and consequently more efficiently analyzed.

Any discussion of embryonic induction will necessarily have to lean heavily upon criteria of embryonic differentiation, and before discussing methods of analyzing embryonic induction and differentiation in molecular terms, it might be useful to restate briefly the various levels of induction and differentiation the embryologist encounters.

Although a rigid classification of these phenomena according to "levels" would be invalid because of the extreme variability and overlap, rough categories are possible. The first may be considered as the differentiation of the zygote and the regionalization occurring during cleavage. These restricted areas in themselves offer enormous difficulties for analysis, and will not be considered here. A second level would be the stimulation of neurulation by the underlying chorda mesoderm. The agent involved, if indeed an agent is involved, in evoking this differentiation response (neurulation) has been aptly called the "organizer" (Spemann, 1927), and the event is frequently thought of as being the primary event of embryonic induction. From these inductive and differentiating events we pass to the third level of induction and differentiation, the morphogenetic differentiation of the embryo. This is the most diverse level. Following neurulation in the vertebrate embryo, mesodermal somites are blocked out, mesenchyme takes up various positions and patterns, the endoderm undergoes a plethora of evaginations and mesenchymal interactions, the embryonic nephros rapidly grows caudad, the appendages form, and many morphological structures take shape. Concomitant with these events a fourth level is reached, the acquiring of specific and advanced metabolic patterns in the late embryo and the early adult (cf. Flexner et al., 1960; Nemeth, 1962).

All of these levels of induction and differentiation afford a continuum of development involving many instances of both gross and subtle interactions. The many secondary inductions and subsequent differentiations of the fourth level may be considered specific responses creating specific metabolic patterns, which in turn are the result of previous interactions. The continuum is then extended in an interminable cycle of growth and development.

The scope of this paper is concerned with experiments pertaining to levels three and four. Three examples of induction and/or differentiation will be given to further amplify the embryologist's concern with these problems.

I. Experimental modification of degrees of differentiation manifested by tissue dissociation and short-term reaggregation.

II. An apparently anomalous instance of chondrogenesis in explants of embryonic mesonephros.

III. A more detailed account of the interactions involved in the induction of vertebral cartilage.

I. Experimental Modification of Degrees of Differentiation Manifested by Tissue Dissociation and Short-Term Reaggregation

The metabolic patterns imparting a differentiated state upon a cell may be morphologically indistinct and exhibited only by the behavior of the cell. This is exemplified by some reaggregation patterns of dissociated cells. For example, the larval salamander epidermis has a very characteristic heterogenous cellular construction. Between two layers of epithelial cells (the external squamous layer and the layer apposed to the basement lamella) lie glandular cells, interstitial epidermal cells, and various types of pigment cells (melanophores, guanophores, etc.). All of these cell types are associated with one another in a regular pattern according to the species characteristics. After dissociation, however, they reaggregate not into a heterogenous pattern again, but into homogenous clumps. Glandular Leydig cells adhere only to cells of the same type, guanophores recombine only with guanophores, etc. (Figs. 1–3). Morphologically they appear unaltered in the sense that cell types are still recognizable, but as evidenced by their behavior they are in some manner changed. *In vivo* they possess a certain level of differentiation which maintains a regular heterogenous pattern. Upon tryptic dissociation this level of differentiation is lost, and the cells exhibit another level of differentiation by regrouping into homogenous cell clumps. If these cells had

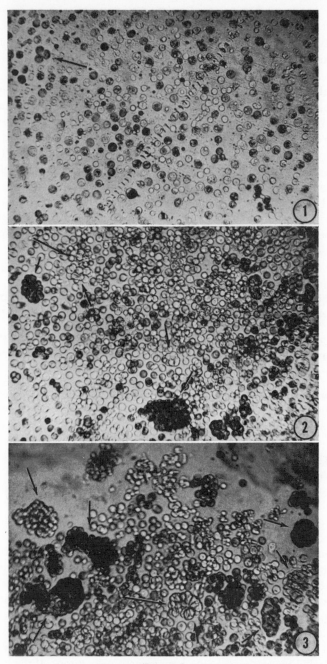

Figs. 1–3.

238

remained unaltered they would be expected to regroup according to the heterogenous patterns shown by others for dissociated chick tissues (Moscona, 1960). It is also possible that to regroup in a specific heterogenous pattern, the cells have to undergo a stimulation akin to induction. For the salamander epidermis, it is likely that during a heterogenous organization the cells utilize their metabolic machinery for a specific type of adhesion and the secretion of the characteristic vast quantities of mucus. When using their metabolic machinery for an irregular task, however, such as a different selective adhesion, this might prevent heterogenous regrouping and the production of mucus. Casual observations indicated that when the cells were allowed to stick to a glass-fluid interface, cell type adhesion did not occur, and a weak attempt at heterogeneity took place. The cells were not followed long enough for adequate observations on the mucus composition of the ambient fluid.

Differentiated states have frequently been studied by cell morphology, e.g., tubular formation in metanephric anlagen (Grobstein, 1954), cartilage formation in somite explants (Holtzer, 1961, review), or lymphocyte production in thymus explants (Auerbach, 1961a, b). There is, however, no real division between cellular morphology and cellular activity and the morphological approach is increasingly leading into the study of metabolic or biosynthetic activity. Since qualities of morphology and behavior are reflections of metabolic or biosynthetic activity, the physical appearance of cells is an indispensable index of differentiation used in conjunction with the increasing body of biochemical knowledge.

II. An Apparently Anomalous Instance of Chondrogenesis in Explants of Embryonic Mesonephros

The characteristic biochemical and morphological properties of cartilage afford suitable parameters for analyzing embryonic chondrogenesis. The chondrogenesis associated with explanted mesonephric tissue gives a good example of the complexities involved in studying embryonic systems.

Figs. 1–3. *Salamandra maculosa* larval epidermis, dissociated with tryptic digestion (3% trypsin in Holtfreter's solution for 20 minutes). Dissociated cells transferred to Holtfreter's solution and photographed at intervals. Magnification, 45×.

Fig. 1. Cell types seen immediately after dissociation: pigment cells, glandular cells, epidermal cells. Three pigment cells forming homogenous clump are indicated by arrow.

Fig. 2. Reaggregation after 15 minutes. Homogenous cell clumps indicated by arrow. Cell triad indicated by arrow in Fig. 1 is shown in upper left by the short arrow.

Fig. 3. Reaggregation after 75 minutes. Homogenous cell groups larger and cells show strong adherence for neighboring cells.

The chick mesonephros begins to form after about 50 hours of incubation, developing tubules associated with the pronephric duct (now called the mesonephric duct) at the level of the 10th somite; and the whole system of duct and tubules grows caudad until it reaches the level of the 30th somite during the third day of incubation. This embryonic kidney becomes functional in the embryo, but is replaced by the metanephric kidney in the post-hatched chick. The mesonephric duct remains for part of its extent as a genital duct, whereas the tubules become nonfunctional and disappear except for those associated with the genital structures.

Except for the fact that the mesonephros lies between two potential cartilage-forming areas (somite and limb bud), it normally has no obvious bearing upon chondrogenesis *in situ*. If, however, the mesonephros from a 3-day-old chick embryo is removed and placed in culture, it will invariably form cartilage (Figs. 4–6). The cartilage appears on the third or fourth day of culture, with a variable persistence of mesonephric tubules. The tissue when placed in culture consists of duct, tubules, and mesenchyme, and the chondrogenesis that occurs during the culture period is not encountered in normal development.

It was thought that one cellular element, e.g., the interstitial mesenchyme, might be transient to the vicinity of the kidney tubules and eventually reach a normal cartilage-producing area to become incorporated into somitic, limb-girdle, or rib-sternum cartilages. All of these cartilage-forming areas are in close proximity to the developing kidney. If the chondrogenic cells are of a transient nature, then they would be expected to form cartilage if dissociated from the other nephric cells.

When this dissociation was done, either manually or by tryptic digestion, an interesting example of tissue interaction was observed in subsequent cultures. The mesenchyme alone did not form cartilage in culture, whether removed by manual dissection or by tryptic digestion. The manually trimmed duct and tubules did form cartilage, whereas the trypsin dissociated tubules and duct did not do so. This suggested that neither the tubules themselves nor the mesenchyme alone were capable of cartilage formation, which indeed seemed to be the case, since cartilage was produced when the dissociated mesenchyme was recombined with the tubules and duct. If these chondrogenic cells are not then transient, it becomes difficult to classify this *in vitro* phenomenon embryologically. The results cited indicate either a transformation of kidney cells into cartilage cells, or a specific tissue interaction similar to other epithelial-mesenchymal systems which result in a particular path of differentiation.

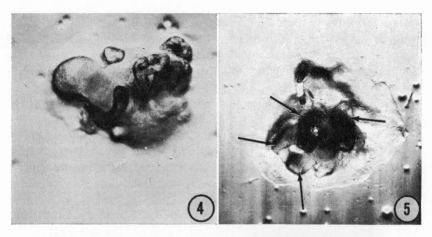

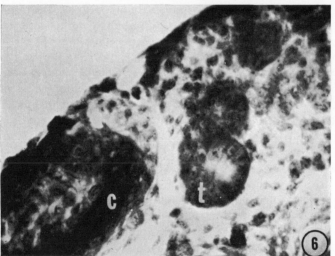

FIG. 4. Stage-18 limb bud grown on nutrient agar for 10 days in association with strip of mesonephros. The mesonephros (stage 17) was from abdominal region. One nodule of cartilage seen to left, epithelial tubules shown at left. Living culture, magnification, 16×.

FIG. 5. Stage-18 limb bud grown on nutrient agar for 10 days in association with strip of mesonephros. The mesonephros (stage 17) was from limb bud region. Multiple rods of cartilage indicated by arrows. Epithelial tubules not present. Living culture, magnification, 16×.

FIG. 6. Histological section of manually trimmed stage-17 mesonephros, 6 days on nutrient agar. Cartilage (c) seen in close proximity to persisting tubules (t). Frequently, however, tubules do not persist. Stained with aqueous thionin, magnification, 430×.

Since the nephric tubules and duct do not always persist in cartilage-forming cultures, cell transformations cannot be discounted. In histological preparations of progressively aged cultures some instances of tubule disorganization were observed, but other cases showed the persistence of well-organized tubules and even swollen ducts.

One clue as to the biological merits of these observations comes from attempts to determine which areas of the kidney are chondrogenic. The mesonephros extends almost the entire length of the thoraco-abdominal region. If the ends of the kidney (i.e., adjacent to the anterior and posterior limb anlagen) are dissected and placed in culture, cartilage formation always occurs. The kidney from the intervening abdominal region either does not form cartilage, or forms very small nodules. That this chondrogenic activity occurs only in these regions strongly suggests that it is pertinent in some way to either limb or girdle cartilage. To check this point, the mesonephros was grown in association with limb buds in culture. Limb buds from 3-day incubates, explanted to nutrient agar, will normally form a single nodule of cartilage, with little evidence of morphogenetic advancement as far as limb bone structure is concerned. The mesonephros from the limb bud region had a marked enhancing effect both upon limb bud chondrogenesis and morphogenesis, whereas the mesonephros from the intervening abdominal area had little or no effect (Figs. 4 and 5). The cartilage that formed in the enhanced cultures took the shape of multiple rods or of one enlarged irregular cartilage mass (Lash, 1963).

In summary, certain cells associated with the embryonic mesonephros are capable of undergoing chondrogenesis in an explant situation. The regions that form cartilage imply that this phenomenon is in some way related to limb or girdle formation. Two lines of evidence indicate that this chondrogenesis is due to a tissue interaction and not the inadvertent inclusion of adjacent chondrogenic areas: (a) dissociation experiments show that neither the tubules nor the immediate mesenchyme are capable of forming cartilage alone, but will do so in recombination, and (b) neither the adjacent somites (cf. Holtzer, 1961) nor the adjacent lateral plate tissue can autonomously form significant amounts of cartilage if explanted from the embryo at the same age as in the mesonephros experiments. Somites from stage-19 embryos (68–72-hour incubation) will regularly form cartilage under nutrient agar culture conditions, and the lateral plate issue will not do so until after stage 16 (51–56-hour incubation) (N. Lash, unpublished). Embryonic mesonephros will form cartilage consistently from the earliest stages at which it can be decently dissected (stages 14–15, 50–55-hour incubation).

The evidence at present would seem to indicate that the chondrogenic cells associated with the mesonephros eventually contribute to the formation of girdle or limb cartilages. There is no *in vivo* evidence for this yet, only the recombination experiments with limb buds. In spite of the circumstantial evidence *in vitro*, we cannot discount the possibility that these chondrogenic transformations are created by the artificial experimental conditions, and may have little developmental significance.

III. A More Detailed Account of the Interactions Involved in the Induction of Vertebral Cartilage

A chondrogenic system which has been analyzed in greater detail is that of vertebral cartilage induction (cf. Holtzer, 1961). For all vertebrate embryos thus far studied, the formation of vertebral cartilage has been found to be specifically related to the inductive influence of the embryonic spinal cord and notochord.

The analysis of the tissue interactions involved in the induction of vertebral cartilage may be summarized as follows:

(1) *In vivo*, the removal of the spinal cord or notochord at an early stage of embryonic development will impair or inhibit the formation of vertebral cartilage. Conversely, extra cartilages will be induced to form in lateral somitic tissue by the implantation of embryonic spinal cord (Holtzer and Detwiler, 1953).

(2) *In vitro*, somites can be taken from an embryo at an early stage of development so they will not form cartilage in culture. Cartilage will form if these somites are cultured with spinal cord or notochord.

(3) The cartilage that is induced to form *in vitro* appears on the fourth day of culture. This interval is quite constant, but may be altered under certain conditions (cf. Stockdale *et al.*, 1961).

(4) These tissue interactions are developmentally specific. No other tissues from these embryos are capable of reacting with the somites in a manner stimulating chondrogenesis, and only the somites will respond to the inducing tissues. In addition the ventral portion of the spinal cord possesses inducing abilities, not the dorsal portion. This is considered a specific tissue induction since no other tissue of developmental significance (i.e., that might be effective during normal development) has been found to affect the behavior of the somites in culture. Even the embryonic mesonephros, capable of autonomous chondrogenesis, has no observable effect upon the contiguous somites.

The interval between induction and the appearance of cartilage is undoubtedly the most important as far as mechanisms of induction and differentiation are concerned, but as yet we have very little information about this period. Transfilter experiments, with the spinal cord on one side and the somites on the other, have indicated that the phenomenon of induction takes place during the first few hours, even though cartilage does not appear until much later (Lash *et al.*, 1957). Once the somites have been stimulated, the continued presence of the inducing tissue is no longer necessary.

This is understandable when it is considered that the somites have passed through stages of differentiation before experimentation. They are, in a sense, prepared to respond to a chondrogenic stimulus of the spinal cord or notochord. This stimulus cannot be utilized by other tissues, potentially chondrogenic or not. So at one level of induction the somites are made reactive at an early stage of development, whereas at a later period they need the added influence of the spinal cord or notochord to fulfill their chondrogenic potentialities. There is yet no evidence as to whether different tissues are active during early stages of induction, nor is there any evidence about how many stages of induction may be recognized. For our discussions, we are using the term induction as it applies to the last known interaction before chondrogenesis ensues.

The analysis of some biochemical events attending the induction of vertebral cartilage falls into three major categories:

A. The appearance of specific macromolecules (viz., chondromucoprotein) in relation to induction and morphological differentiation.

B. Attempts at simulating the molecular events attending the inductive event.

C. Clues to metabolic patterns initiated or stimulated by the inductive event.

A. *Specific Macromolecules*

Chondroitinsulfate is a characteristic extraction product of cartilage, the matrix of which is primarily chondromucoprotein (Malawista and Schubert, 1958; Muir, 1961) and collagen. Neither the collagen nor the protein is suitable for assay procedures since they are not characteristic enough of cartilage, although recent work demonstrating a s-RNA–hydroxyproline complex may be useful in the analysis of cartilage-collagen biosynthesis (Manner and Gould, 1962). Whereas the chondroitinsulfates are not unique to cartilage matrix, for our experimental conditions they are specific enough to be used in bioassay procedures.

If induced-somite cultures are grown in the presence of inorganic radioactive sulfate the extracted chondroitinsulfate is radioactive due to the incorporation of the labeled sulfate, and it was found that measurements of the radioactivity in a chondroitin sulfate extract were reasonable, although not accurate determinations of the amount of cartilage synthesized in cultures (Lash *et al.*, 1960).

This procedure was used to determine when the characteristic constituents of cartilage appeared after the somites had been induced. Within the limits of the procedures used, chondroitin sulfate was not detectable prior to the visible appearance of matrix on the fourth day of culture (Fig. 7). Thus, during the interval between induction and visible histogenesis, there is no significant build-up of macromolecular constituents. Since the inductive event occurs within the first few hours, the metabolic response to induction must involve biosynthetic steps prior to the synthesis of the macromolecular complex, chondromucoprotein.

B. *Simulating the Inductive Event*

Another approach followed was the analysis of the initiation of chondro-genesis, i.e., the inductive event as it occurs *in vitro* with 3-day-old embryonic tissues. This work has been reported in greater detail elsewhere (Lash *et al.*, 1962), and will be reviewed here.

The unavoidable inference from past work is that an agent (or agents) from the spinal cord and notochord are transmitted to the somite cells, there stimulating the final differentiating steps toward chondrogenesis. This induction fits the criterion of all inductive events in that it is an interaction, or communication, between two dissimilar tissues. Attempts at characteriz-ing a transmissible agent by interposing a porous membrane between the spinal cord and somites have in the past proven uneventful (Lash *et al.* 1957), but recently more promising results have been obtained with ex-traction procedures (Lash *et al.*, 1962, also cf. Strudel, 1962).

The rationale behind the extraction procedures was that nucleotides, either as coenzymes or complexes, might be involved with either the inductive

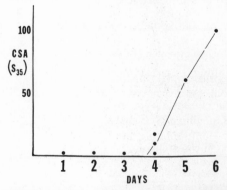

FIG. 7. Formation of radioactive chondroitinsulfate (CSA), extracted from notochord-induced somites. Amount formed in 6-day-old cultures arbitrarily set as 100. (Details in Lash *et al.*, 1960.)

event or the metabolic patterns induced within the somite cells. Nucleotides, as coenzymes or related compounds, uridine triphosphate (UTP), uridine diphosphate (UDP)-glucose, UDP-N-acetylhexosamine, or as PAPS, the "active sulfate principle" (3'-phosphoadenosine 5'-phosphosulfate—Robbins and Lipmann, 1955, 1958; Suzuki and Strominger, 1960), play a very prominent role in the biosynthesis of polysaccharides (Boström, 1959; Roseman, 1959; Kent, 1961). The rationale may have been erroneous, but the extraction procedure was fruitful.

Various combinations of the inducing tissues (spinal cord, notochord) and reacting tissues (somites) were grown in the presence of inorganic radioactive phosphate, and the labeled tissues were then extracted in cold perchloric acid (0.25 M) for acid-soluble nucleotides. After passing the perchloric acid extract over a charcoal-celite column, the sugar phosphates pass through and a nucleotide eluate is obtained by eluting with 10% aqueous pyridine.

The perchloric acid extract or the nucleotide eluate from 3-day-old spinal cords or notochords was found to be chondrogenically active. When added to reactive somites, the incidence of cartilage formation was comparable to that obtained with intact inducing tissues (Table I). Although an adequate

TABLE I

CARTILAGE FORMATION IN CHICK SOMITES AFTER THE ADDITION OF EXTRACTS OBTAINED FROM SPINAL CORDS AND NOTOCHORDS[a]

	Number of cases	Cartilage
Whole extracts or eluates (3- and 4-day embryo)	172	132
Controls (no extracts or control extracts)	379	65
Fractionated eluate (chondrogenic fraction) (4-day embryo)	170	134
Eluate minus chondrogenic fraction (4-day chick)	118	16
Other tissue extracts tested		
Spinal cord, ventral (3-day embryo)	30	23
Spinal cord, dorsal (3-day embryo)	30	3
Limb muscle (5-day embryo)	10	4
Epidermis (3-day embryo)	10	2
Vertebral cartilage (7-day embryo)	10	2
Mesonephros (3-day embryo)	19	2
Somites (stage-18 embryo)	10	6
Liver (10-day embryo)	20	10

[a] Cumulative figures, somites of stages 16, 17, 18.

number of experiments have not yet been performed, extracts from other tissues tested did not possess comparable activity (Table I). Unfortunately quantitative data on the amount of extract or eluate necessary are not yet available, but the quantity was very small for these experiments. Considering the amount of embryonic spinal cords or notochords used, the amount of extract must have been less than 1 μg. This extract was dissolved in 1.0 ml of saline, and mixed with 1.0 ml of horse serum and 0.5 ml of embryo extract (11-day chick). A drop of this nutrient medium (0.25 ml) added to the somite clusters induced cartilage to appear on the fourth day of culture, as in the tissue-induced explants.

Table I shows some of the results obtained, although the number of cases for some instances is small enough to make the results tentative. Attempts at isolating a chondrogenic factor from 4-day embryos indicate that there is a component in the cold perchloric acid extract that is capable of stimulating chondrogenesis in explanted somites (Fig. 8). The chemical characterization of a biologically active component has not yet been completed (Hommes *et al.*, 1962). This is being performed on 4-day embryos, whereas the bulk of the biological experiments are done with 3-day embryos. The problems encountered in this type of analysis are numerous, particularly if the aim is to obtain evidence for a specific chemical induction.

Considering the complexities of the many events of induction and differ-

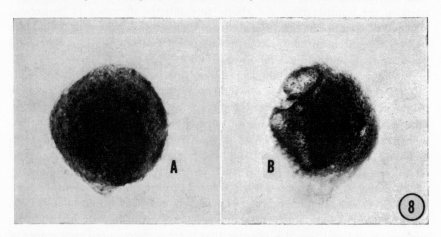

Fig. 8. Clusters of 6 stage-17 somites grown on nutrient agar for 6 days. Culture A received nutrient medium; culture B received nutrient medium containing charcoal eluate of cold perchloric acid extract obtained from 4-day-old embryonic chick spinal cords and notochords. Two of the five cartilage nodules present are prominent on left side of culture. Living cultures, magnification, 16$\times$.

entiation, and the possibility of chemical differences between 3- and 4-day-old tissues, it would be unfortunate, but not too surprising, to find that the results already obtained may come to be categorized as nonspecific stimulation, and not specific induction. Even a cursory review of the embryological literature would show that the characterization of chemical "inducers" is a formidable task (cf. Needham, 1950; Niu, 1959; Yamada, 1961).

Strudel (1962) has recently reported that a saline extract of the spinal cord is capable of inducing cartilage formation in explanted somites. It is not known whether the saline extract and the perchloric acid extract share a common active component.

It is yet too early to state categorically that the events stimulated by the nucleotide extract mimic the inductive event and have biological reality. With these qualifications in mind, we are trying to determine whether these observations have developmental significance, and at the same time are attempting to analyze the metabolic response of the somite cells. The biological validity and specificity of these experiments will come forth in the future, but clues to the metabolic response are presently available.

C. *Metabolic Response*

At the time of explantation, the 3-day-old somites are passing through progressive stages of differentiation, with unknown inductions associated with the differentiating events. The induction under consideration here imparts to the somitic tissue the ability to fulfill the direction of its normal course, i.e., chondrogenesis. The ability of some of the somite cells to undergo myogenesis has been considered in papers by Holtzer (1961, review) and his co-workers and will not be discussed here.

Previously the only metabolic response discernible in the somitic tissue was the appearance of cartilage matrix. Other metabolic events must of necessity precede the formation of the complex cartilage matrix. Staining progressively aged induced cultures with Azure B Bromide and for ribonucleic acid (RNA) (Swift, 1955) indicated that there is a progressive increase in the amount of RNA basophilic material in the tissues. This could have been surmised beforehand, since protein synthesis must undoubtedly play a major role in any metabolic response resulting in the biosynthesis of a complicated extracellular matrix. This protein synthesis would include both the production of ribonucleoproteins and necessary enzymes. Although no evidence is at hand yet, it is possible that there is a sequential appearance of specific coenzymes (e.g., UTP, uridine triphosphate), specific enzymes (e.g., epimerases), and metabolic intermediates (e.g., N-acetylgalactosamine). At one stage in this sequence, the influence of the inducing tissues is necessary to establish the complete machinery for matrix biosynthesis.

Initial experiments approaching this problem were done in the following manner. Since it was known that ester-bound polysaccharide sulfate was not very prominent until matrix formation (Lash, *et al.* 1960), evidence was sought for other types of sulfate fixation during the interval between induction and visible differentiation. Induced somites (somites plus spinal cord or notochord) were grown in the presence of radioactive sulfate as for the earlier experiments (Lash *et al.*, 1960), and fixed at intervals for the preparation of autoradiographs. It was seen that the inducing tissues themselves incorporated radioactive sulfate as well as the noninduced somites. The degree of incorporation was initially slightly greater in the inducing tissues, but the incorporation in the somites was markedly enhanced as early as 4 hours after contact with the inducing tissues (Figs. 9 and 10). The fixation of sulfate by the somites continued progressively during the first few days, increasing in the amount seen in autoradiographs (Figs. 9–13). At the time of matrix formation, a cartilage nodule was quite prominent in autoradiographs (Figs. 13 and 22). Therefore, one of the metabolic responses is the fixation of sulfate in increasing quantities prior to matrix formation, after which the sulfate is then extractable as chondroitinsulfate. When autoradiographs of the early stages of sulfate fixation were extracted for polysaccharides, there was no detectable diminishment of the autoradiograph pattern (Fig. 14). Polysaccharide extractions performed on tissues possessing cartilage showed a marked decrease in grain pattern (Figs. 16 and 23). This indicated that the sulfate fixed in the early stages of differentiation is not in the form of ester-bound polysaccharide sulfate.

To determine further the manner in which sulfate was bound by chondrogenic tissue, autoradiographs of sectioned cultures possessing cartilage matrix were extracted with the enzyme ribonuclease (RNase) (Fig. 17). In these experiments there was a marked diminution of the autoradiograph pattern. It cannot be said at present that the extractable sulfate is ribonucleoprotein sulfate, since there exists the possibility that the ribonuclease preparation* might have been contaminated with other proteolytic enzymes. The RNase-sensitive sulfate binding was most evident in the interval prior to visible differentiation, whereas the polysaccharide extractable sulfate was most prevalent after cartilage matrix was formed.

These two types of sulfate fixation (RNase-sensitive and polysaccharide) are accompanied by at least one more type, the "active sulfate principle" (PAPS) (Robbins and Lipmann, 1955). Although PAPS is extractable from cultures possessing cartilage, this analysis has not yet been carried back to the early stages after induction. A sequential pattern of the various types of

* The ribonuclease preparation was obtained from Nutritional Biochemical Corporation.

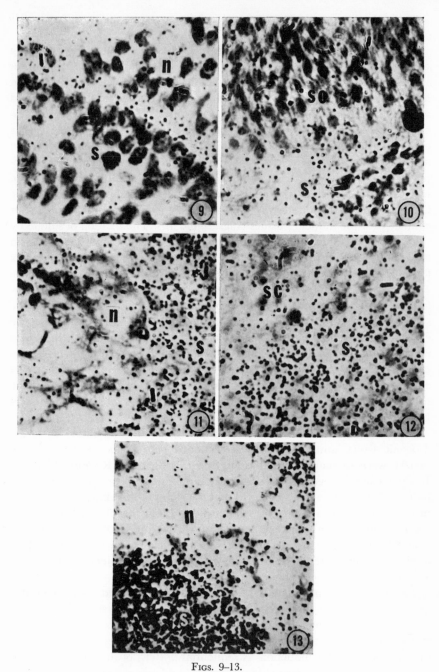

FIGS. 9–13.

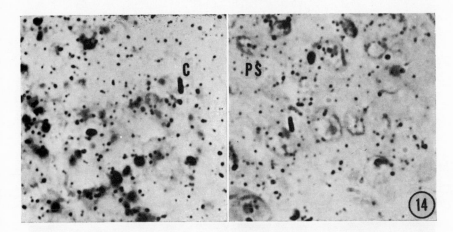

FIG. 14. Somite cultures grown for 6 hours on nutrient agar. Sections of tissue on left (*c*) extracted with water for 72 hours; sections of tissues on right (*ps*) extracted for polysaccharides with 30% KCl and 2% K₂CO₃ for 72 hours. Extraction for polysaccharides did not diminish components responsible for sulfate incorporation. Untreated autoradiographs had the same appearance as the two photographs shown. Methods and magnification same as in Figs. 9–13.

sulfate fixation has not yet been obtained, but ancillary experiments indicate its possible importance.

(1) Embryonic mesonephros cultures incorporate sulfate during the first few days of culture. This fixation is most prominent in mesonephros from

FIGS. 9–13. Autoradiographs of inorganic sulfate incorporation in notochord-somite and spinal cord-somite combinations. Tissues grown in presence of Na₂S³⁵O₄. Tissues stained with aqueous thionin, focused on silver grains in emulsion. Symbols are *n*, notochord; *sc*, spinal cord; *s*, somites. Magnification, 970×.

FIG. 9. Notochord-somite combination, 4 hours in culture. Initial incorporation in region of notochordal sheath, with slight incorporation both in notochord and in somites. Oriented cells are somitic cells.

FIG. 10. Spinal cord-somite combination, 4 hours in culture. Initial incorporation in somitic cells next to spinal cord tissue. Incorporation in both tissues, but slightly more in somites.

FIG. 11. Notochord-somite combination, 72 hours in culture. Noticeable increase in the incorporation of sulfate by somitic cells. Incorporation in notochord is primarily in cytoplasmic processes, not in vacuoles.

FIG. 12. Spinal cord-somite combination, 72 hours. Comparable increase of sulfate incorporation to that in Fig. 10.

FIG. 13. Notochord-somite combination, 5 days in culture. Heavy incorporation in cartilage in lower left and extreme upper right corners. Incorporation in notochord has not changed appreciably from 4-hour cultures.

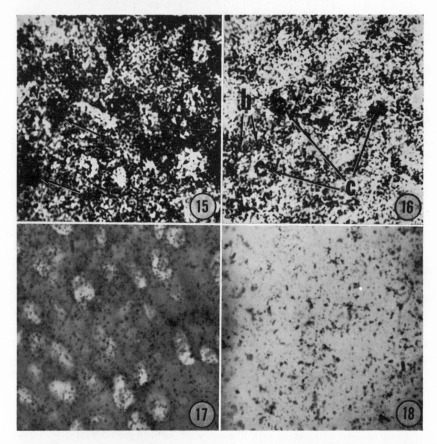

FIGS. 15–18. Autoradiographs of 7-day chick vertebral cartilage exposed to inorganic radioactive sulfate. Sections stained with aqueous thionin, photographs focused on silver grains. Magnification, 970×.

FIG. 15. Untreated cartilage section. Intense incorporation except for few lacunae not possessing cells. Where the cartilage cell is present in section, it is obscured by grains (*arrows*). Tissue strongly metachromatic, but obscured by grains.

FIG. 16. Sections extracted for polysaccharides as in Fig. 14 (*ps*). Appreciable amount of sulfate incorporation still present, mostly in the cartilage cells (*c*) in the section, and in the "halo" (*h*) around lacunae. Sections show no metachromasia and stain lightly.

FIG. 17. Sections extracted with RNase (0.2 mg/ml at pH 6.5 for 2 hours at room temperature—Swift, 1955). Marked decrease in components responsible for sulfate incorporation, primarily in the region of the cartilage cells, which are prominent in sections. Some incorporation seen in matrix, which exhibits prominent metachromasia.

FIG. 18. Sections extracted with RNase, then extracted for polysaccharides. Bulk of sulfate is removed, as are all staining properties. Residual incorporation seen may indicate incomplete polysaccharide extraction.

the limb bud area and negligible in that from the abdominal area. This sulfate is resistant to polysaccharide extraction, but is diminished with RNase treatment. It has not yet been possible to localize the fixation pattern to a particular cellular element, e.g., mesenchyme or tubules.

(2) Extractions of mature vertebral cartilage indicate the presence of three

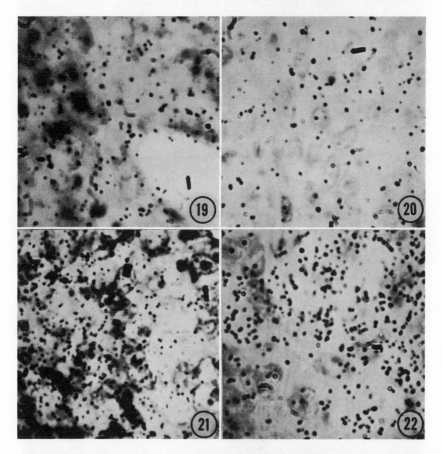

FIGS. 19–22. Autoradiographs of stage-17 somite cultures, exposed to inorganic radioactive sulfate. Histological methods and magnification same as in Figs. 9–13.

FIG. 19. 24-hour culture, receiving nutrient medium only.

FIG. 20. 72-hour culture, receiving nutrient medium only.

FIG. 21. 24-hour culture, receiving nutrient medium containing cold perchloric acid extract from embryonic spinal cords and notochord.

FIG. 22. 72-hour culture, receiving nutrient medium containing cold perchloric acid extract as in Fig. 21.

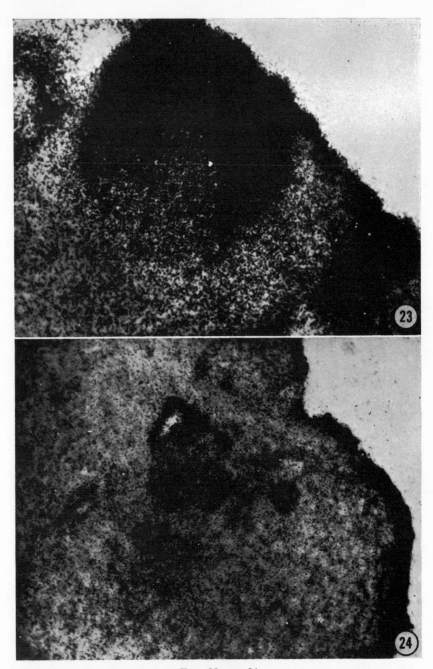

Figs. 23 and 24.

types of sulfate fixation: PAPS, polysaccharide, and RNase-sensitive. The RNase-sensitive sulfate is primarily associated with the nucleus and/or cytoplasm (cf. Figs. 15–18), whereas the polysaccharide sulfate is primarily associated with the matrix. Practically all sulfate fixation is removed with both extractions (Fig. 18). The halo of sulfate fixation seen around the lacunae (Fig. 16) is suggestive of the site where sulfate transfer takes place, but no critical evidence of this has been obtained.

(3) The stimulation of sulfate fixation is initiated in somitic tissue both by the inducing tissues and by a perchloric acid extract obtained from the inducing tissues (Figs. 19–24). Whereas explanted somites will normally fix sulfate to a slight degree, fixation is stimulated by the inductive event. Contrastingly, the fixation by chondrogenic mesonephros proceeds without the intervention of an external agent or tissue. The implication of these facts is that the ability to incorporate sulfate into an RNase-sensitive molecule may be one of the determining factors for chondrogenic tissues.

(4) Work by Chapeville and Fromageot (1957) supports the interpretation that the autoradiograph patterns reflect sulfate fixation, and not its reduction products (sulfite, sulfhydryl groups). The body tissues in the embryonic chick apparently lack the enzymes necessary for sulfate reduction (also cf. Lowe and Roberts, 1955; Machlin et al., 1955).

In summary, one metabolic event initiated in induced somites is the incorporation of sulfate into an RNase-sensitive molecule. This fixation proceeds until the time of visible differentiation, when the sulfate is presumably transferred onto the polysaccharide molecule. This information should prove useful in the analysis of the metabolic patterns attending the induced chondrogenic differentiation.

Discussion

The criterion of induction is an interaction or communication between cells, and the criterion of response is the differentiation of specific metabolic patterns. As there are different levels of induction, so are there different levels of differentiation, all blending into a complex continuum of embryonic growth and development. Failure to give full cognizance to these different levels has led to an unnatural reliance upon morphological criteria. For its

---

Figs. 23 and 24. Autoradiographs of stage-17 somites exposed to inorganic radioactive sulfate. Culture received cold perchloric acid extract obtained from embryonic spinal cords and notochords, grown on nutrient agar for 5 days. Magnification, 430×.

Fig. 23. Nodule of cartilage clearly visible, showing heavy sulfate incorporation.

Fig. 24. Same tissue extracted for polysaccharides, cartilage nodule no longer visible. Large amount of nonextracted sulfate still present.

era, morphological differentiation was sufficient for study, but with the availability of more exacting assay methods, it ceases to be wholly satisfactory.

The differentiation of a cell is an infra-morphological phenomenon. Before a cell produces muscle proteins or the mucoproteins of connective tissue ground substance, it must acquire the necessary metabolic machinery. The problem of embryonic (and postembryonic) induction and differentiation concerns the methods by which cells acquire and maintain these specific metabolic patterns (cf. Holtzer *et al.*, 1960).

Because of the complexities involved with embryonic induction and differentiation, it is understandably difficult to obtain an all-encompassing concept of embryology. The terms induction and differentiation are used in common by immunologists, microbial geneticists, and experimental embryologists. In the former two disciplines the terms are well prescribed and have definite parameters; and induction is acknowledged as a molecular interaction. A reacting system is induced to yield a specific response by synthesizing a particular enzyme, antibody, etc. Most embryologists would like to think of embryonic induction as a molecular event, inducing a specific metabolic response (i.e., differentiation) and some are attempting to define embryological problems in these terms. Since the embryologist has little hope at present of analyzing induction in molecular genetic terms as the microbiologists have done, the most suitable approach at present seems to be to work back from the differentiated state of metabolic activity to its initiation.

The seeming universality of inductive and differentiative phenomena in the realm of biology raises interesting points as to its teleonomic necessity. During the evolution of multicellular organisms and the division of cellular labor attending this evolution, it became important that the cells be able to communicate in some way with other cells (cf. Weiss, 1955). It is an apparent intimate communication for the lower metazoans and for some of the tissue interactions involved during embryogenesis. In some cases, however, notably the development of germ cells, thymus lymphocytes, and hormonal glands, it became necessary to utilize humoral communication. In this sense humoral communication, e.g. hormonal action, is an obvious extension of a generalized concept of induction and differentiation (cf. Monod and Jacob, 1961).

The preparation of reacting tissues to respond to inductive stimuli is yet another order of discussion, and one in which very little information is available. It is undoubtedly axiomatic, however, that all reacting tissues studied by the immunologist, microbial geneticist, endocrinologist, and embryologist have been primed to respond in a particular fashion to the inductive stimulus. At any particular point in the continuum, it is safe to say that there was a preceding interaction. The inductive events studied by most researchers involve the elicitation of a definite and limited response. In many cases the differentiating response has been shown to have a genetic basis (cf., Ingram, 1958; Davis, 1961), and this basis is theorized in other instances.

In this sense all instances of induction are equivalent, whether they involve the induction of an immunological response, a morphogenetic response, or a hormonal response. The action of a hormone is in many ways no different from the action of an embryonic inducing tissue. The major difference is in the distance the agent is transmitted and the chemical knowledge we have of the (inducing) hormonal agent. It may be that some types of hormonal action are inductive events *par excellence*, and some aspects of embryonic induction involve processes similar to hormonal induction. Even embryonic systems involving so-called "pluripotent" reacting tissues are equational to the systems discussed in this paper. The complexities of histogenetic and morphogenetic differentiation are so little known that at present it is difficult to see how any one scheme of induction will suffice, if couched in too specific a terminology (cf. Weiss, 1962). A conceptual definition comparable to hormonal action may be quite useful, at least it may serve a remedial purpose to the current trend toward extreme reductionism (Simpson, 1962).

Such a concept also offers to the embryologist a useful adjunct to the more specific genetic-molecular concept currently in vogue. The embryologist is not now equipped, nor is his material well enough known in most instances, to give information related to genetic action in the strict sense that the microbiologist uses the term induction.

One final point may be made, viz., it is one thing to know the conditions whereby the stimulation of a specific biosynthesis is effected, but this is still a bit removed from specific tissue and organ morphogenesis. Neural arches form in one chondrogenic system, and ribs, long bones, or girdles in another; one group of muscle cells combine to form the anterior tibialis and another group the soleus. Physical conditions play an important role (cf. Weiss, 1947), but this is an unsatisfactory and partial answer. The problems of cell behavior and organization will acquire greater importance as the embryologist and the biochemist approach the problems of induction and differentiation at their various levels.

Acknowledgments

The author is Helen Hay Whitney Established Investigator, Department of Anatomy, School of Medicine, University of Pennsylvania, Philadelphia, Pennsylvania. The investigation was supported by grants A-1980 from the United States Public Health Service and G-14123 from the National Science Foundation. The author gratefully acknowledges the counsel of Drs. H. Holtzer, M. W. Whitehouse, F. Zilliken, and others who have contributed in many ways to the results reported in this paper.

REFERENCES

AUERBACH, R. (1961a). Experimental analysis of the origin of cell types in the development of the mouse thymus. *Develop. Biol.* **3**, 336–354.

AUERBACH, R. (1961b). Genetic control of thymus lymphoid differentiation. *Proc. Natl. Acad. Sci. U.S.* **47**, 1175–1181.

BOSTRÖM, H. (1959). Biosynthesis of sulfated polysaccharides. *Trans. Josiah Macy Jr. Conf. on Polysaccharides in Biol., 4th Conf.* pp. 211–275.

CHAPEVILLE, F., AND FROMAGEOT, P. (1957). Formation de sulfite, d'acide cystéique et de taurine à partir de sulfate par l'oeuf embryonné. *Biochim. et Biophys. Acta* **26**, 538–558.

COHEN, S. (1954). Virus-induced metabolic transformations and other studies on unbalanced growth. *In* "Aspects of Synthesis and Order in Growth," Growth Symposium No. 13 (D. Rudnick, ed.), pp. 127–148. Princeton Univ. Press, Princeton, New Jersey.

DAVIS, B. D. (1961). The teleonomic significance of biosynthetic control mechanisms. *Cold Spring Harbor Symposia Quant. Biol.* **26**, 1–10.

FLEXNER, L. B., FLEXNER, J. B., ROBERTS, R. B., AND DE LA HABA, G. (1960). Lactic dehydrogenases of the developing cerebral cortex and liver of the mouse and guinea pig. *Develop. Biol.* **2**, 313–328.

GROBSTEIN, C. (1954). Tissue interaction in the morphogenesis of mouse embryonic rudiments *in vitro*. *In* "Aspects of Synthesis and Order in Growth," Growth Symposium No. 13 (D. Rudnick, ed.), pp. 233–256. Princeton Univ. Press, Princeton, New Jersey.

HOLTZER, H. (1961). Aspects of chondrogenesis and myogenesis. *In* "Synthesis of Molecular and Cellular Structure," Growth Symposium No. 19 (D. Rudnick, ed.), pp. 35–87. Ronald Press, New York.

HOLTZER, H., AND DETWILER, S. R. (1953). An experimental analysis of the development of the spinal column. III. Induction of skeletogenous cells. *J. Exptl. Zool.* **123**, 335–370.

HOLTZER, H., ABBOTT, J., LASH, J., AND HOLTZER, S. (1960). The loss of phenotypic traits by differentiated cartilage cells. *Proc. Natl. Acad. Sci. U.S.* **46**, 1533–1542.

HOMMES, F. A., VAN LEEUWEN, G., AND ZILLIKEN, F. (1962). Induction of cell differentiation. The isolation of a chondrogenic factor from embryonic chick spinal cords and notochords. *Biochim. et Biophys. Acta* **56**, 320–325.

INGRAM, V. M. (1958). Abnormal human haemoglobins. I. The comparison of normal human and sickle-cell haemoglobins by "fingerprinting." *Biochim. et Biophys. Acta* **28**, 539–545.

JACOB, F., AND MONOD, J. (1961). Genetic regulatory mechanisms in the synthesis of proteins. *J. Mol. Biol.* **3**, 318–356.

KENT, P. W. (1961). Biochemistry of sulphated mucosubstances. *Biochem. Soc. Symposia (Cambridge, Engl.)* **20**, 90–108.

LASH, J. (1963). Studies on the ability of embryonic mesonephros explants to form cartilage. *Develop. Biol.* **6**, 219–232.

LASH, J., HOLTZER, S., AND HOLTZER, H. (1957). An experimental analysis of the development of the spinal column. VI. Aspects of cartilage induction. *Exptl. Cell Research* **13**, 292–303.

LASH, J. W., HOLTZER, H., AND WHITEHOUSE, M. W. (1960). *In vitro* studies on chondrogenesis: the uptake of radioactive sulfate during cartilage induction. *Develop. Biol.* **2**, 76–89.

LASH, J. W., HOMMES, F. A., AND ZILLIKEN, F. (1962). Induction of cell differentiation. The *in vitro* induction of vertebral cartilage with a low-molecular-weight tissue component. *Biochim. et Biophys. Acta* **56**, 313–319.

LOWE, I. P., AND ROBERTS, E. (1955). Incorporation of radioactive sulfate sulfur into taurine and other substances in the chick embryo. *J. Biol. Chem.* **212**, 477–483.

MACHLIN, L. J., PEARSON, P. B., AND DENTON, C. A. (1955). The utilization of sulfate sulfur for the synthesis of taurine in the developing chick embryo. *J. Biol. Chem.* **212**, 469–475.

MALAWISTA, I., AND SCHUBERT, M. (1958). Chondromucoprotein; new extraction method and alkaline degradation. *J. Biol. Chem.* **230**, 535–544.

MANNER, G., AND GOULD, B. S. (1962). Collagen biosynthesis. The formation of an s-RNA-hydroxyproline complex. *Federation Proc.* **21**, 169.

MOSCONA, A. A. (1960). Patterns and mechanisms of tissue reconstruction from dissociated cells. *In* "Developing Cell Systems and Their Control," Growth Symposium No. 18 (D. Rudnick, ed.), pp. 45–70. Ronald Press, New York.

MONOD, J., AND JACOB, F. (1961). Teleonomic mechanism in cellular metabolism, growth, and differentiation. *Cold Spring Harbor Symposia Quant. Biol.* **25**, 389–401.

MUIR, H. (1961). Chondroitin sulphates and sulphated polysaccharides of connective tissue. *Biochem. Soc. Symposia (Cambridge, Engl.)* **20**, 4–22.

NEEDHAM, J. (1950). "Biochemistry and Morphogenesis." Cambridge Univ. Press, London and New York.

NEMETH, A. N. (1962). The effect of 5-fluorouracil on the developmental and adaptive formation of tryptophan pyrrolase. *J. Biol. Chem.* **237**, 3703–3706.

NIU, M. C. (1959). Current evidence concerning chemical inducers. *In* "Evolution of Nervous Control," pp. 7–30. Am. Assoc. Advance. Sci., Washington, D.C.

ROBBINS, P. W., AND LIPMANN, F. (1955). Isolation and identification of active sulfate. *J. Biol. Chem.* **229**, 837–851.

ROBBINS, P. W., AND LIPMANN, F. (1958). Separation of the two enzymatic phases in active sulfate synthesis. *J. Biol. Chem.* **233**, 681–685.

ROSEMAN, S. (1959). Metabolism of connective tissue. *Ann. Rev. Biochem.* **28**, 545–578.

SIMPSON, G. (1962). The status of the study of organisms. *Am. Scientist* **50**, 36–45.

SPEMANN, H. (1927). Organisers in animal development. *Proc. Roy. Soc. (London)* **B102**, 177–187.

SPEMANN, H. (1938). "Embryonic Development and Induction." Yale Univ. Press, New Haven, Connecticut.

SPEMANN, H., AND MANGOLD, H. (1924). Über Induktion von Embryonalanlagen durch Implantation artfremder Organisturen. *Arch. mikroskop. Anat. u. Entwicklungsmech.* **100**, 599–638.

STOCKDALE, F., HOLTZER, H., AND LASH, J. (1961). An experimental analysis of the development of the spinal column. VII. Response of dissociated somite cells. *Acta Embryol. Morphol. Exptl.* **4**, 40–46.

STRUDEL, G. (1962). Induction de cartilage *in vitro* par l'extrait de tube nerveux et de chorde de l'embryon de Poulet. *Develop. Biol.* **4,** 67–86.

SUZUKI, S., AND STROMINGER, J. L. (1960). Enzymatic sulfation of mucopolysaccharides in hen oviduct. *J. Biol. Chem.* **235,** 257–276.

SWIFT, H. (1955). Cytochemical techniques for nucleic acids. *In* "The Nucleic Acids" (E. Chargaff and J. N. Davidson, eds.), Vol. 2, pp. 51–92. Academic Press, New York.

WEISS, P. (1947). The problem of specificity in growth and development. *Yale J. Biol. and Med.* **19,** 255–278.

WEISS, P. (1955). Specificity in growth control. *In* "Biological Specificity and Growth," Growth Symposium No. 12 (E. G. Butler, ed.), pp. 195–206. Princeton Univ. Press, Princeton, New Jersey.

WEISS, P. (1962). From cell to molecule. *In* "The Molecular Control of Cellular Activity" (J. M. Allen, ed.), pp. 1–72. McGraw-Hill, New York.

YAMADA, T. (1961). A chemical approach to the problem of the organizer. *In* "Advances in Morphogenesis" (M. Abercrombie, ed.), Vol. 1, pp. 1–53. Academic Press, New York.

Author Index

Numbers in italics show the page on which the complete reference is listed.

Subject Index

A

Acetabularia, 209, 164
Acetobacter xylinum, 215
Acetylpyridine, effect on protein nitrogen, 96
Active sulfate principle, 246, 249
Actomyosin, 106
Aging, 181
Alleles of regulator genes for *E. coli*, 39
Allosteric
  enzymes, 33
  inhibition, 30
  protein, 51
Allosteric effects, specificity of, 31, 32
Amino acid activation in chick leg muscle, 106, 107
Amino acid substitution, 15, 22, 23
  relationship to genetic code, 22
σ-Aminolevulinic acid, 158, 160, 162, 166
β-Aminopropionitrile, 185
Amyloplast, 169
Amyloplastids, 144
Anisotropy of growth, 209
Ascaris, 53
Aspartic transcarbamylase, 32
Avena, 209

B

Bacteria, *see* also individual organisms
  lysogenic, 38
  regulation of protein synthesis in, 35, 36
Balbiani ring, 120, *see* also Puff
Barley, 147, 156, 157, 158, 161
Beans, effect of red light on, 157
Biosynthesis of polysaccharides, 246

C

Carbohydrates in chloroplasts, 148
Carotenes, 158
Carotenoids, 144
  in chloroplasts, 148
Cartilage
  formation, *see* also Chondrogenesis
  in chick somites, 246
  in mesonephros explants, 241

nodules, 247
  vertebral, induction of, 243
Cathepsin C, 192
Cell walls
  optical properties of, 213
  synthesis of, 208
  texture in, 210
Cellular differentiation, 30
Chaetomorpha, 215, 228
Chara, 210
Chick embryos, 186
  explants, 88 ff
  incorporation of phenylalanine and fluorophenylalanine, 93
  increase in protein nitrogen, 90
  lathyritic skeletons of, 186
  protein synthesis in explants, 88
Chick leg muscle, 107, 108
  RNA in, 108
  phospholipid in, 108
  protein nitrogen in, 108
Chironomus, 58, 59, 120, 125, 126, 128, 130, 135
  chromosomes, puffing and ecdysone, 136
Chlamydomonas, 164
Chloramphenicol, 92
Chloroplasts
  CO_2 fixation in stroma, 150
  development, 154
  O_2 production from, 150
  orientation in algae, 158
Chlorella, 158
Chlorophyll, 144 ff, *see* also Plastid and Chloroplast
  biosynthesis in developing plastid, 158, et seq.
Chloroplasts, 144 ff
  analysis of, 147, 148
  lamellar structure of, 151
  separation from mitochondria, 149
  structure, 144
    and photosynthetic function, 149
Chloroplast files, 218, 219
Chondrogenesis, 237
  in embryonic mesonephros, 239
Chondrogenic factor, 247

269